Artificial Intelligence and Machine Learning

This book focuses on artificial intelligence (AI) and machine learning (ML) technologies and how they are progressively being incorporated into a wide range of products, including consumer gadgets, "smart" personal assistants, cutting-edge medical diagnostic systems, and quantum computing systems. This concise reference book offers a broad overview of the most important trends and discusses how these trends and technologies are being created and employed in the applications in which they are being used.

Artificial Intelligence and Machine Learning: An Intelligent Perspective of Emerging Technologies offers a broad package involving the incubation of AI and ML with various emerging technologies such as Internet of Things (IoT), healthcare, smart cities, robotics, and more. The book discusses various data collection and data transformation techniques and also maps the legal and ethical issues of data-driven e-healthcare systems while covering possible ways to resolve them. The book explores different techniques on how AI can be used to create better virtual reality experiences and deals with the techniques and possible ways to merge the power of AI and IoT to create smart home appliances.

With contributions from experts in the field, this reference book is useful to healthcare professionals, researchers, and students of industrial engineering, systems engineering, biomedical, computer science, electronics, and communications engineering.

Innovations in Health Informatics and Healthcare: Using Artificial Intelligence and Smart Computing

Series Editors: Rashmi Agrawal, Manav Rachna International Institute of Research and Studies, and Mamta Mittal, G.B. Pant Government Engineering College

The aim of this series is to publish reference books and handbooks that will provide conceptual and advanced reference material centered around health informatics and healthcare using AI and smart computing. There are numerous fields within the healthcare sector where these technologies are applicable including successful ways of handling patients during a pandemic. Large volumes of data, data analysis, and smart computing devices like IoT for sensing health data have drastically changed the way the healthcare sector functions. The scope of the book series is to report the latest advances and developments in the field of health informatics with the use of the latest technologies. Each book will describe in detail the use of *AI, Smart Computing, Evolutionary Computing, Deep Learning, and Data Analysis in the field of Health Informatics*, and the books will include real-life problems that focus on the healthcare system.

Intelligent Computing Applications for COVID-19: Predictions, Diagnosis, and Prevention
Edited by Tanzila Saba and Amjad Rehman

Blockchain for Healthcare Systems: Challenges, Privacy, and Securing of Data
Edited by Shiekh Mohammad Idrees, Paural Agarwal, and M. Afshar Alam

Healthcare Systems and Health Informatics: Using Internet of Things
Edited by Pawan Singh Mehra, Lalit Mohan Goyal, Arvind Dagur, and Anshu Kumar Dwivedi

For more information on this series, please visit: https://www.routledge.com/Innovations-in-Health-Informatics-and-Healthcare-Using-Artificial-Intelligence-and-Smart-Computing/book-series/CRCIHIHUAISM

Artificial Intelligence and Machine Learning

An Intelligent Perspective of Emerging Technologies

Edited by
Rohit Tanwar
Surbhi Bhatia Khan
Varun Sapra
Neelu Jyoti Ahuja

CRC Press is an imprint of the
Taylor & Francis Group, an **informa** business

First edition published 2024
by CRC Press
2385 NW Executive Center Drive, Suite 320, Boca Raton FL 33431

and by CRC Press
4 Park Square, Milton Park, Abingdon, Oxon, OX14 4RN

CRC Press is an imprint of Taylor & Francis Group, LLC

ISBN: 978-1-032-47846-3 (hbk)
ISBN: 978-1-032-48294-1 (pbk)
ISBN: 978-1-003-38831-9 (ebk)

DOI: 10.1201/9781003388319

Typeset in Times
by KnowledgeWorks Global Ltd.

Contents

Editors

Rohit Tanwar earned a bachelor's degree and PhD in computer science engineering at Kurukshetra University, Kurukshetra, India, and a master's degree in computer science engineering at YMCA University of Science and Technology, Faridabad, India. He has more than 10 years of experience in teaching and is currently working as an Associate Professor in the School of Computer Science at the University of Petroleum and Energy Studies (UPES), Dehradun, Misraspatti, India. Dr. Tanwar has more than 40 publications to his credit in books and reputed journals and conferences. His areas of interest include network security, optimization techniques, human computing, soft computing, cloud computing, data mining, etc.

Surbhi Bhatia Khan holds a doctorate in computer science and engineering in machine learning and social media analytics. She earned a Project Management Professional Certification at the reputed Project Management Institute, Newtown Square, Pennsylvania, USA. She is currently working as a Lecturer in the Department of Data Science, School of Science, Engineering and Environment, University of Salford, Manchester, United Kingdom. She has more than 11 years of academic and teaching experience in different universities. She has published 100+ papers in many reputed journals in high indexed outlets. She has around 12 international patents from India, Australia, and the United States. She has authored and edited 12 books. Her areas of interests include information systems, sentiment analysis, machine learning, databases, and data science.

Varun Sapra is presently associated with the School of Computer Science, University of Petroleum and Energy Studies, Dehradun, Uttarakhand, India. He earned a PhD in computer science and engineering at Jagannath University, Jaipur, Rajasthan, India. His 17 years of experience is a blend of both industry and academia. Before joining academia, he was in the corporate sector and worked at Cupid Software, Web Opac Applications, CMA, and many more. He has contributed more than 25 publications in peer-reviewed international journals and conferences and has published 5 patents. His research areas include machine learning, decision support systems, case-based reasoning, and self-organizing maps.

Neelu Jyothi Ahuja is a Professor and Head of the Department of Systemics at the School of Computer Science at the University of Petroleum and Energy Studies, Dehradun, Uttarakhand, India. She earned a PhD in 2010 and has conducted government-sponsored research and development projects and consultancies. Dr. Ahuja has more than 22 years of experience in teaching, research, and project proposal development, and she has published papers in journals and conferences at the international and national levels. Her areas of interest include teaching-learning conversations, mentoring activities, and holistic student engagement and experience development. She strongly believes in learning to be the first step toward teaching and teaching beyond formal classroom boundaries.

Editors

1 Deep Learning Strategies in Biomedicine Imaging Technique

P. Jayadharshini, Santhiya S, Keerthika S, and Priyanka S
Kongu Engineering College (Autonomous), Perundurai, Erode, Tamil Nadu, India

INTRODUCTION

Modern medicine makes use of natural phenomena like light, radiant energy, radioactivity, nuclear magnetic resonance, and sound to produce visualizations or images of internal or external parts of the human body or a portion of the human body either quasi or through an invasive procedure. X-ray radiography, computed tomography (CT) scans, magnetic resonance imaging (MRI) scans, echocardiogram, and virtual pathology are the simulations that are most frequently used in biomedicine. One of the most significant sources of evidence for medical intervention and clinical analysis is imaging data, which makes up about 90% of all healthcare data in total.

The structure and characteristics of internal organs can also be examined by clinicians using scientific imaging without the need for any invasive procedures. Today, a wide range of different picture types are used in the field of research imaging. These modalities use non-invasive procedures as opposed to conventional radiography, which uses x-rays to give the radiologist a viewpoint on the structural and functional state of tissues. As imaging technology advances and becomes more helpful, more tests could be performed to assess vascular changes, compositions, and blood absorption. There are many different types of diagnostic imaging, including CT, radar, and positron emission tomographic (PET) imaging.

In the public's field of vision, deep learning (DL) has grown since its emergence as a subfield of machine learning (ML) in 2006. It is a process control technique that makes use of multiple computational levels made up of numerous linear models or many layers of complex structures [1]. In the fields of informatics, ML, speech synthesis, computational linguistics, and audio identification, supervised learning has recently achieved strides. Because of the vast potential for its use in data analysis, since 2013, one of the top 10 breakthroughs has been named as training data. Tchapga et al. described the connecting layers with artificial neurons, and the artificial neural network (ANN), one of the ML algorithms, was created [2]. It was inspired by the functioning of the human brain system. However, overfitting and

DOI: 10.1201/9781003388319-1

vanishing gradient issues have plagued ANN for training deep networks due to the low computing power and lack of learnable data. When it comes to computer vision and speech recognition tasks, deep neural networks (NNs) outperform human or other ML capabilities due to the increase of computing power with graphics processing units and the availability of big data acquisition.

Medical imaging has greatly advanced in contemporary medicine and is now capable of taking a variety of biomedical images of patients [3]. Then, using these photos and training and intelligent system to analyze them, medical professionals can utilize them to identify the many ailments that can be detected. In this context, classification is crucial because it helps to organize these photos into categories of diseases and increases the subsequent stage of a computer-aided diagnosis system. In ML, the idea of categorization addresses the issue of determining to which category or categories a new population belongs. An observation-filled training set of data is used to classify data when the membership of a category is known. To classify biomedical images, this chapter will survey existing classification techniques. To apply these methods to a big data architecture, the system shows how to leverage the Spark framework. Additionally, it suggests a classification methodology based on the discovered best algorithms, including DL and support vector machine, which were taken from the literature. The suggested classification workflow includes a number of customizable steps, all of which use the same feature extraction approach that is shown in this chapter.

One area of artificial intelligence (AI) that has grown rapidly in recent years is DL [4]. The scientific world has focused on DL due to its versatility, high performance, strong capacity for generalization, and diverse applications, among many other features. The spread of more powerful computers and the amount of medical data have fueled even more interest in this area. Starting with the most fundamental concept and incorporating the mathematical models that support this type of AI's functionality, it gives an overview of current DL methodologies. First, the basic idea behind ANNs is presented, followed by discussions of convolutional structures, recurrent networks, attention models, and finally the transformer, which is the most recent structure. The second step is the introduction of all the fundamental training concepts and other standard architectural design components. Next, several of the essential components of contemporary networks for segmenting and classifying medical images are displayed. The primary DL-related aspects are then highlighted in a discussion of some recent applications that were developed. The ideas and expectations for DL's future are offered in the end. DL is used by simulating the human NN. Multiple nonlinear processing layers are combined to abstract the data from scratch, layer by layer. The data are then processed to extract various levels of abstract properties that are used for target segmentation, categorization, or identification.

EMERGING TECHNIQUES OF MEDICAL INSTRUMENTS

Biomedical inventions like the microscope, x-ray, UV rays, and CT scan are described in Figure 1.1. The evolution from microscope to MRI scan images helps to diagnose critical diseases efficiently.

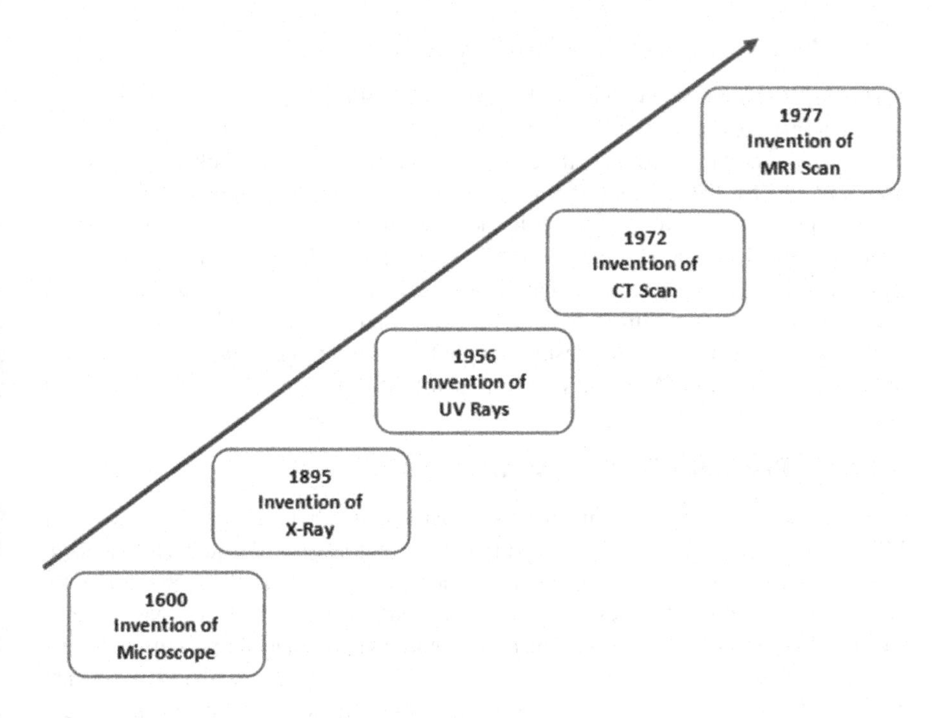

FIGURE 1.1 Evolution of medical tools.

Computed Tomography (CT)

The process of autonomously acquiring x-ray images in which a patient is exposed to a focused beam of x-rays that is swiftly spun around the body is known as "computerized tomography." This generates signals, which the computer's computer subsequently processes to create bridge scans, or "servings," of the organ. These slices are called tomographic images, and they offer more information than regular x-rays. To build multi-images of the patient, the machine's computer may automatically "layer" a number of subsequent slices together. When the camera system's one-dimensional (1D) images are all finished rotating, it appears to form a series of overlay sinusoidal images. Sonograms are the name for these unprocessed diagnostic imaging data, and they are the most often used source in enterprises nationwide. Finally, an iterative reconstruction methodology is used to turn the ultrasound data into a tomographic image of the patient's inner organs.

Positron Emission Tomography (PET)

PET is a form of nuclear anticancer operation that monitors the energy usage of the cells in healthy cells. In reality, it combines nuclear medicine with laboratory analysis. It aids in the visualization of biochemical changes occurring in the organ, like heart body's respiration. PET is mostly used in patients with cancer, brain, or heart disorders. Other nuclear diagnosis tests do not detect metabolism within tissue cells, making it a unique test to measure how much radioactive material has accumulated in a particular area of body tissue.

Single-Photon Emission Computed Tomography (SPECT)

A single-photon emission computed tomography (SPECT) test is an integrated test that demonstrates how blood flows to cells and organs. It is used to detect convulsions, accidents, broken bones, illnesses, and cervical tumors. Nuclear echocardiograms like SPECT combine imaging tests with a nuclear emitter. The tracer allows medical professionals to study how blood flows to cells and organs. Your blood is injected with a tracer before the SPECT scan. The radiolabeled tracer generates gamma rays that the CT scanner can detect because it is radiolabeled. The gamma rays' information is captured by the computer and shown on CT cross sections. Combining these methods can result in a photorealistic representation of your brain in these cross sections. As you lay on a table, the ASPECT scanner spins around you.

Magnetic Resonance Imaging (MRI)

Anatomical images in three dimensions (quasi visualization) are produced using an MRI. Disease detection, diagnosis, and therapy monitoring are frequently done using this process. It is based on reducing electronics that excite and reflect changes in the axis of revolution of particles in the organ accordion water. Massive magnetism is employed in MRIs to produce a powerful gravitational pull that compels the cell's protons to match with the magnetism. After that, the patient is pulsed with radiofrequency energy, which stimulates the protons and causes them to spin out of balance and struggle against the magnetic field. The protons realign with the magnetic field when the microwave field is switched off, creating energy that the MRI sensors can detect. The surroundings and chemical makeup of the molecules affect how long it takes for the protons to realign with the magnetic field and how much energy is released. These magnetic characteristics allow doctors to distinguish between different kinds of tissue.

Optical Coherence Tomography (OCT)

Optical coherence tomography (OCT), a recently created imaging method, has been extensively used in the domain of biomedical scanning. Recently, dermatology, cardiology, and ophthalmology have all found uses for it as a diagnostic tool. The primary subjects of this chapter are applications in the domains of glaucoma and ocular mapping. OCT images have speckle noise, a granular pattern caused by the underlying physics, which limits the interpretation process. To get rid of the noise and keep the visual features, special noise reduction techniques are needed. An important development in the study of OCT images is the outcome of feature extraction for distinguishing between various features, particularly those that are multifocal. OCT collects thickness maps of the various retinal layers, the process' typical output, which are particularly helpful for studying both healthy and sick patients. The emergence of pathology in the region under surveillance, as well as changes in the organism, have an impact on the quality and accurate interpretation of the recorded pictures, necessitating the employment of various image registration procedures. To process image data and convert it into a format that professionals can easily understand, a variety of techniques are currently used, which are examined in this chapter.

DEEP LEARNING APPROACHES

Numerous algorithms, each with its own benefits and drawbacks, exist in the popular research area of DL. These techniques practically cover every aspect of our image representation, with a primary emphasis on segmentation and classification.

Classification

DL for image categorization is a growing field that has seen early success. The most common structure among them is the convolutional neural network (CNN). DL technique has experienced an explosion [5], which was built on the DL model CNN, and these techniques practically cover every aspect of our image representation. The inter-network structure was introduced by Lin et al. [6] and to decrease the risk of collinearity, it makes use of average global pooling [7]. In 2014, the ImageNet dataset's accuracy was increased by both Google Neural Network and VGGNet [8]. Versions v2, v3, and v4 of GoogleNet have also been developed to improve productivity [9]. He et al. [4] created the spherical gradient accumulating to increase the robustness of the supplied data in response to CNN's shortcomings with the predetermined criteria for input vector. They also formulated the residual network Deep residual to address the issue of potential model deterioration as the trained model got more complex and to advance the field's industrial advancement. Kim et al. [10] present a concise exploration of the application of deep learning in medical imaging within the context of neurospine-related studies. Their review underscores the transformative potential of deep learning techniques for enhancing image analysis and diagnosis, fostering advancements in neurospine healthcare. Simonyan and Zisserman [11] introduce very deep convolutional networks, pioneering the architecture's efficacy for large-scale image recognition tasks. Their paper laid the foundation for modern deep convolutional neural networks, revolutionizing computer vision and enabling breakthroughs in image classification.

Consider the case of AlexNet. Multilayers and fully connected layers make up an eight-layer network topology that the AlexNet adopted in 2012. A maximum pooling is carried out to lessen the amount of data after each of the five NNs' convolution layers. AlexNet will accept input data up to 227 × 227 pixels. the 66256 feature matrix was eventually submitted to the completely interconnected layer following five variations of convolution and pooled techniques. In the fully interconnected layer's sixth layer, 4,096 convolution parameters are set up, along with a 4,096-size linear feature value that was obtained during the dropout procedure. The mistake rate for AlexNet in ImageNet was 15.3%, which was substantially higher than the error rate for the runner-up at 26.2%. While doing so, it adopted ReLU as its activation function, demonstrating that it is more effective than sigmoid activation.

Segmentation

DL's feature extraction area of study is significant. As DL technology advances quickly, several superior semantic segmentation NNs appear and continue to advance to the top of many segmentation competitions. People began to use CNN for picture segmentation as a result of its success in the classification sector. CNN may accept image data of any resolution; however, due to layers being fully connected at the channel's end, it will miss important information when grouping for collecting attributes and geographical details from the raw image. CNN finds it challenging to identify the category to which particular pixels belong. Some segmentation networks with a convolutional structure have been developed as transfer learning technology progresses.

The first semantic distributional networks were created by Long et al. [7] using the fully convolutional network (FCN). It accomplishes pixel-level classification by substituting convolutional layers for the classification network VGG16's fully connected layers while keeping the feature map's spatial information. The image is then restored by FCN utilizing deconvolution and feature maps, and it then offers the segmentation outcome for each pixel via softmax. The multilayer perceptron, which has dense connections, is replaced by the fully connected layer; this greatly reduces the amount of features that need to be trained because it is directly connected and exchanges information. In comparison to the previous approach, the FCN's performance on the Pascal VOC 2012 datasets [3] increased by 20%, achieving 62.2% of the performance.

V-Net, which was developed on the basis of FCN, has been extensively applied in medical imaging. To restore the dimension and quality of the image, V-Net creates an audio codec structure for feature extraction utilizing the concept of FCN texture analysis. By repeatedly combining the layers to gather the data samples, the encoder gradually reduces the high spatial resolution, and the decoder portion slightly increases the dimension and recovers the desired detail based on the feature data.

DEEP LEARNING ALGORITHMS

CNN

The most popular DL architecture is CNN because of its resemblance with the traditional NN. In contrast to a standard NN, CNN works with source images and a multi-network of neurons that only link to a fraction of the preceding stage rather than the entire layer. Convolutional, pooling, or fully connected layers are among the layers that make up the CNN. Repaired regular layers are among the nonlinear transfer layers (ReLUs) that are nonlinear activation layers. Volumes of extracted features containing the features the filter retrieved needed to be created; the fully connected layer conducts convolution between the input picture's components and the filters. In the function *fx* 14 max0, *x* is used by ReLU, an activation functions layer, to alter the input values.

Because they are based on neighboring pixels and descending the component values to reduce the computational expense and spatial density of the image, pooled stack computations are translation invariant. The final layer of a CNN is typically a convolution because all of its neurons are coupled to those in the layer before it, making it similar to the hidden layers of a regular NN. CNN is frequently employed to overcome categorization issues, as was previously indicated. For CNN image retrieval, the input image is divided into tiny patches of the same size. This NN has awarded a classification to the patch's pixel value. The following center pixel is then classified as the patch moves forward. However, this method is ineffective because it does not reuse the moving patterns' matching features, which results in the loss of spatial information once the features become the many ultimate fully connected nodes. The completely connected convolutional network was suggested as a solution to this issue, wherein the inverted segmentation process was switched to on for the last fully connected layers of the CNN, which simultaneously perform feature extraction and apply upsampling to the lower-resolution extracted features to recover the original spatial size.

Deep NNs are often trained using a training methodology technique in conjunction with an optimization technique like gradient descent. The technique entails calculating the gradient of an error function to update the network weights and minimizing the loss feature value. A previous study [12] highlights the significance of deep learning techniques in enhancing diagnostic accuracy and disease detection across various medical specialties. By analyzing recent advancements and methodologies, the review offer valuable insights into the potential of deep learning for revolutionizing medical image analysis.

Restricted Boltzmann Machines (RBMs)

Restricted Boltzmann machines (RBMs) are multilayer NNs that draw inspiration from energy-based models. The relationships between the variables are encoded by allocating scalar energy to every arrangement of the variables. Finding the remaining independent variables in an energy-efficient way requires interpretation or inference based on the observed data values. Finding an analytical equation that requires lower values for the appropriate value of that related to the presence and higher energies for improper combinations is how learning is performed. The loss function, which is minimized during learning, offers a gauge of the efficacy of the available energy functions.

Autoencoder-Based Deep Learning Architectures

The input is compressed into a dormant description using a fully convolutional computational model, an unconstrained process of learning, using a backpropagation algorithm technique with predicted values that are similar to the inputs. It is split into two parts: (1) the input is compressed by the encoding into a dormant form, which is demonstrated by that of the component $h = f(x)$, and (2) the input is then decoded and rebuilt by the decoding from the latent spatial field. By forcing the concealed gradient's coefficients to be less than those of the NN, compression is achieved. Such a network can be described as inadequate. The network learns the most prominent characteristics in the learning set because of the reduced dimensionality of the hidden layer.

Architectures for Deep Learning Based on Sparse Coding

Sparse coding, a kind of transfer classification, chooses an excessively comprehensive number of basis vectors to represent the input data. When a hidden depiction is over complete, its dimension exceeds that of the input. The goal is to find a linear function of these column vectors that matches a specific input. Due to the over completeness of the network, extra constraints on sparsity must be used to deal with any degeneracy. Sparse coding has the advantage of being able to both find correlations between descriptors of similarity and capture significant aspects of images.

Recurrent Neural Networks

Recurrent neural networks (RNNs) are specially made to handle entries from measuring the time taken when the size of the input cannot be known in advance. These networks, like RNNs, rely on both the present input and knowledge gained from prior values. The network's hidden state vector contains the previous input data. This means that the same source may yield various results depending on the earlier inputs in the series. Repeatedly changing the network with different power phase vectors generated from various input series values causes the network to start repeating. The concealed state is refreshed following every input.

DATA AUGMENTATION

Having access to enough data is necessary for DL NNs to function effectively. The problem is that, especially in the field of diagnostic imaging, training samples are frequently not accessible in adequate numbers. A variety of data augmentation techniques are used to create new instructional data from an existing dataset to enhance the information provided in the deletion of accurate data. The data augmentation procedure involves applying category preserving changes to the image data and could involve (1) changing the orientation of image pixels without altering the resulting image dimensions, (2) flipping rows and columns of pixels to create toggling the image both horizontally and vertically, (3) rotating the image by a predetermined amount between 0 and 360 degrees, (4) adjusting the image intensity to enable the model to take these variations in experimental data into account, and (5) randomly linear interpolation or adding more border pixels to enlarge or shrink the image. Most of these approaches effectively remove some of the existing images and replace them with fresh ones, either via nearby filling, replicating border boundaries, blending, or extrapolation.

The primary goal of these augmentation techniques is to broaden the deep NN platform. Usually, during the network's training phase, these strategies are employed automatically. A safer method to utilize is linear transformations, which are also adequate in the majority of situations.

Heavy augmentations, on the other hand, could result in unnaturally varied characteristics. The final decision, however, will once more be heavily based on the characteristics of the target region and the type of medical imaging. Last but not least, although data augmentation techniques can be employed to address the data scarcity, they are unable to account for all potential variations in the raw data.

SOFT COMPUTING

PSPNet provides a pooling hierarchical system to aggregate the collected information under multiscale on the same level. By establishing data fragmentation, this

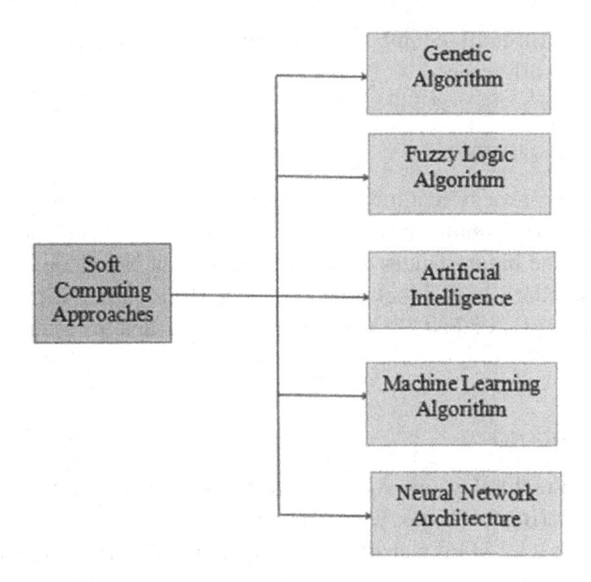

FIGURE 1.2 Approaches of soft computing.

structure solves the issue of how scenes and global information relate to the testing environment, which FCN cannot effectively handle. The capacity to gather global information is improved by its pooling pyramid structure, which may aggregate context information from several places. Several significant soft computing technologies are included in this review [13] including evolutionary algorithms, ANN, and fuzzy logic (FL) algorithms. Hard computing is not as accepting of uncertainty, inaccuracy, incomplete truth, and approximations as soft computing techniques. When given more latitude to carry out their duties within the confines of their position, they perform better. Due of its adaptability and precision, scientists suggest and employ soft computing technology the most. It can also solve difficult problems and is advantageous because it is economical, effective, and efficient.

Soft computing approaches like genetic algorithm, NN architecture, ML algorithm, AI, and FL algorithm are shown in Figure 1.2. These algorithms are effectively applied to accurately diagnose from the medical image, which helps medical practitioners work toward a quick treatment process.

Genetic Algorithm

The principles of genetics and natural selection are the foundation of the optimization algorithm, which is an improved lookup method. It is routinely used to identify flawless or almost flawless answers to challenging issues that would normally require a lifetime to accomplish. This is generally used in ML, research, and optimizing problems. People have always found tremendous inspiration in nature. Genetic algorithm is a lookup algorithm mainly based on the ideas of genetic selection and natural selection. We have a population or pool of potential solutions to the given problem in genetic algorithms. The procedure is continued over a number of generations as these solutions go through recombination and mutation (similar to natural genetics), giving rise to new offspring.

Fuzzy Logic Algorithm

An argumentation strategy that mirrors human reasoning is called fuzzy logic. FL's strategy mimics how humans make decisions, which involve considering all middle options between the binary values of the YES value and NO value. A computer can understand a standard logic block that creates a clear output with accurate input response as TRUE (1) or FALSE (0), which is comparable to a human expressing YES or NO.

Artificial Intelligence

The science and technology of AI form the basis of many disciplines, including computer engineering, physiology, neurology, linguistics, mathematics, and architecture. AI places a lot of emphasis on creating computer algorithms for skills like thinking, learning, and problem-solving that are similar to those of human intelligence. The development of an intelligent system can draw from one or more of the following areas.

Machine Learning Algorithm

The concept of ML was initially brought up in the latter half of the 1980s and the beginning of the 1990s, and machine learning is a term that was first used in 1959 by Arthur Samuel. To carry out certain tasks, a subset of AI enables robot behavior and data-driven decision-making. These systems must be developed in a particular way for them to be gradually exposed to new information so they can learn and develop. There are several applications for ML, including the data taxonomy and clustering, the identification of organs from photographs, and the reconstruction of medical images.

Neural Network Architecture

Particularly, the large, complicated NN-based computational intelligence technique has helped NNs gain a significant amount of attention in recent years. Transfer learning and ANN, two AI-based methods, have developed a novel framework for inverse challenges that could revolutionize the market.

Generative Adversarial Network (GAN)

A constantly changing opposition is a danger. In a standard generative adversarial network (GAN) arrangement, two NNs compete with one another. Two separate GAN systems can be used to generate two different kinds of results. Throughout the strategy, the efficiency of these two networks keeps increasing. The high-pass filter system enables conventional basic NN models, and GAN can handle much more complicated data production issues. Because image processing uses it so regularly, its information skills are frequently exploited. This innovative method makes it easier to perform visual synthesis, image semantic analysis, and design transfer via various networks.

PERFORMANCE METRICS

Standard metrics are used to assess the discrete optimization system's performance that provides comparison with other techniques that have been published. According to the system's functionality, a number of elements must be taken into consideration while choosing an evaluation metric. These measurements could assess, among other things, accuracy, processing speed, memory usage, and computational complexity. The following are some performance measures that can be used to judge how well the DL models perform in terms of segmenting data.

ACCURACY

$$Accuracy = \frac{Correctly\ Predicted\ Pixels}{Total\ number\ of\ Image\ Pixels} = \frac{TP + TN}{TP + FP + FN + TN} \tag{1.1}$$

The percentage of correctly categorized image pixels is known as accuracy, according to the definition given in Eq. (1.1). Despite the fact that it represents the most basic performance measurement, it only accurately represents picture segmentation performance when there is an imbalance in the classes. Class imbalance arises when one segmentation class outperforms the other. In this instance, results will be overstated because the superior precision of the dominating class will obscure the poorer resolution of the opposing class. The accuracy metric should be used to assess segmentation performance using images without any imbalanced data.

PRECISION

$$Precision = \frac{Correctly\ Predicted\ Disease\ Pixels}{Total\ number\ of\ Predicted\ Disease\ Pixels} = \frac{TP}{TP + FP} \tag{1.2}$$

Precision is defined as the percentage of disease particles in the output of image detection that match the underlying pathological images. It is a key indicator of many classifiers due to its vulnerability to over segmentation. The result of having the highest number is low precision scores.

RECALL

$$Recall = \frac{Correctly\ Predicted\ Disease\ Pixels}{Total\ number\ of\ Actual\ Disease\ Pixels} = \frac{TP}{TP + FN} \tag{1.3}$$

Recall is the percentage of genetic disease regions in the ground truth that an ensemble classifier successfully detected. Due to the low recall scores that arise from under segmentation, it is vulnerable to this.

F1 MEASURE

$$F1_{measure} = 2 \times \frac{Precision \times Recall}{Precision + Recall} \tag{1.4}$$

Because both have high values, to get a certain segmentation result, fragmented sections must match the actual data in terms of position, level of detail, precision, and recall. The harmonic average of mean and accuracy is defined by $F1_{\text{measure}}$, also known as Boundary $F1$, and it is useful for matching boundaries or limits between expected and actual segmentation.

DICE SIMILARITY COEFFICIENT (DSC)

$$DICE = \frac{2\left|S_{Ground\,Truth} \cap S_{Automated}\right|}{\left|S_{Ground\,Truth}\right| + \left|S_{Automated}\right|} = \frac{2 \times TP}{2 \times TP + FP + FN} \tag{1.5}$$

In each class, dice similarity coefficient (DSC) takes into account both false alarms and missed values, as can be seen from the equation, making it superior to total spatial precision. Furthermore, DICE is considered to be better because it assesses the correctness of the segmentation boundaries in addition to counting the number of correctly labeled pixels. Furthermore, cross-validation is frequently employed with DICE to gauge the consistency of system performance.

JACCARD SIMILARITY INDEX (JSI)

$$JSI = \frac{S_{Ground\ Truth} \cap S_{Automated}}{S_{Ground\ Truth} \cup S_{Automated}} = \frac{TP}{TP + FP + FN} \tag{1.6}$$

Intersection-over-union, another name for the Jaccard Similarity Index (JSI), is a calculation of how closely two segments match up. Its formula is the ratio of the segments' combined areas when they are projected to match up. Due to their monotonicity or positive correlation, JSI and DSC are comparable.

CONCLUSION

In this chapter on DL techniques for biomedical image segmentation, some significant challenges have been highlighted. The actual data that supported each of these studies demonstrated the suggested method's applicability to the given application using tiny datasets. What is still unclear is why a specific problem can be solved using

DL methodologies. Research is still being done to fully comprehend the response to this query. Modern visual techniques are being created by several professionals to make it easier to comprehend feature maps obtained from concealed nodes. Another difficulty that many researchers overlook is the question of how generalizable the network reaction is when the data source changes, which is what will happen if a data acquisition device changes, as this could result in adjustments to image attributes like lighting or color intensity levels. Network performance will suffer as a result of the lack of generalizability.

Another challenge with DL-based networks is the requirement for very large image datasets. As a result, there will be a significant increase in the networks' training time as well as massive storage and memory requirements. Another area of active research is the reduction of training time as well as the efficient management of memory and storage needs for massive amounts of imaging data. A further obstacle to the development of DL-based techniques for biomedical applications in clinical practice is the absence of suitably big imaging datasets. Despite the fact that the healthcare sector holds a lot of aging-related data, it is not disclosed to the public because it contains protected health information or because the organization views it as a proprietary asset. Data exchange will have protracted advantages that greatly outweigh any gains made by keeping the data secretive in the short term; hence, it is essential that efforts be made to make such data publicly accessible through incentives for main challenges or through information offerings. DL approaches, from segmenting skin lesions to autonomously interpreting CT scans, have enabled unprecedented performance increases in a range of biomedical applications. It is possible to achieve considerably more, though, if more annotated images are made accessible to the general public.

REFERENCES

1. Beutel J, Kundel HL, and Van Metter RL. Handbook of Medical Imaging, vol. 1. SPEI Press, 2000.
2. Tchapga CT, Mih TA, Kouanou AT, Fonzin TF, Fogang PK, Mezatio BA, and Tchiotsop D. Biomedical image classification in a big data architecture using machine learning algorithms. Journal of Healthcare Engineering. 2021;2021:9998819.
3. Everingham M, Van Gool L, Williams CKI, et al. The pascal visual object classes challenge 2012 (voc2012) results. Available online: http://www.pascal-network.org/challenges/VOC/voc2011/workshop/index.
4. He K, Zhang X, Ren S, et al. Spatial pyramid pooling in deep convolutional networks for visual recognition. IEEE Transactions on Pattern Analysis and Machine Intelligence. 2015;37:1904–16.
5. Krizhevsky A, Sutskever I, Hinton GE. ImageNet classification with deep convolutional neural networks. Communications of the ACM, 2012, 6(June 2017), 84–90. doi: 10.1145/3065386.
6. Lin M, Chen Q, Yan S. Network in network. arXiv preprint. arXiv:1312.4400, 2013.
7. Long J, Shelhamer E, Darrell T. Fully convolutional networks for semantic segmentation. In: Proceedings of the IEEE conference on computer vision and pattern recognition. 2015:3431–40.
8. Szegedy C, Liu W, Jia Y, et al. Going deeper with convolutions. In: Proceedings of the IEEE conference on computer vision and pattern recognition. 2015:1–9.

9. Ioffe S, Szegedy C. Batch normalization: accelerating deep network training by reducing internal covariate shift. arXiv preprint. arXiv:1502.03167, 2015.
10. Kim M, Yun J, Cho Y, Shin K, Jang R, Bae HJ, Kim N. Deep learning in medical imaging. Neurospine. 2019 Dec;16(4):657–68. doi: 10.14245/ns.1938396.198.
11. Simonyan K, Zisserman A. Very deep convolutional networks for large-scale image recognition. arXiv preprint. arXiv:1409.1556, 2014.
12. Anaya-Isaza A, Mera-Jiménez L, and Zequera-Diaz M. An overview of deep learning in medical imaging. Informatics in Medicine Unlocked. 2021;26:100723.
13. Kobashi S, Nyúl LG, Udupa JK. Soft computing in medical image processing. Computational and Mathematical Methods in Medicine. 2016;2016:7358162.

2 X-Ray-Based Pneumonia Detection Using ResNet50 and VGG16 Extracted Features and Conventional Machine Learning Algorithms

Amit Virmani, Akhilesh Singh, Ritesh Agarwal, Sanjeet Kumar, and Hemant Kumar
School of Engineering and Technology (UIET),
CSJM University, Kanpur, UP, India

INTRODUCTION

Pneumonia is an infection that affects the lungs and can cause severe respiratory problems. The infection causes inflammation in the lungs, leading to a range of symptoms, including cough, fever, and breathlessness. It is a potentially life-threatening disease caused by bacteria, viruses, or fungi [1]. Pneumonia is a severe and potentially life-threatening condition that can affect people of all ages, particularly young children and older adults. Based on information from the World Health Organization, pneumonia is responsible for a huge number of deaths worldwide [2]. It is a major reason for morbidity and mortality worldwide, and early detection is crucial for timely treatment

One of the most common diagnostic methods for pneumonia is chest x-ray (CXR) imaging, which is used to visualize the lungs and identify any signs of inflammation or infection. One of the significant challenges in pneumonia detection is the difficulty in distinguishing between normal and pneumonia. While normal images on x-ray typically show clear and transparent lung tissue, pneumonia x-ray images may show areas of opacity, fluid accumulation, and inflammation [3].

The difference between normal and pneumonia x-ray images are presented in Figure 2.1 [4]. Figure 2.1a shows a normal CXR, where the lung tissue appears transparent, and there are no signs of inflammation or infection. In contrast, Figure 2.1b shows the CXR of a patient with pneumonia. In this image, the lung tissue appears opaque and hazy, with areas of inflammation and fluid accumulation visible. As these images illustrate, the differences between normal

DOI: 10.1201/9781003388319-2

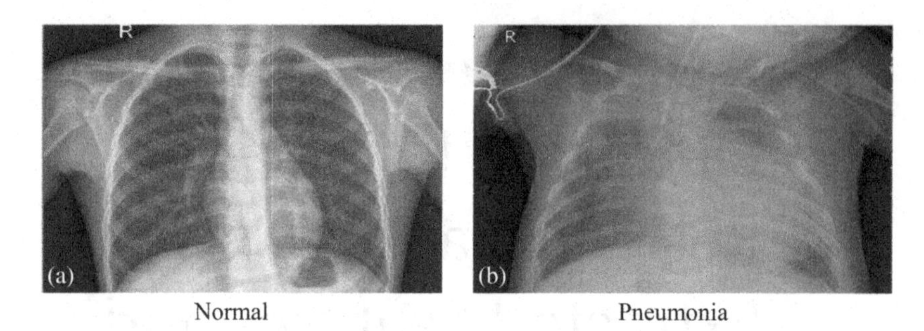

FIGURE 2.1 Instances of x-ray [4].

and pneumonia x-ray images can be subtle, and accurately identifying these differences requires careful analysis and interpretation of the images by trained healthcare professionals.

In recent years, advances in machine learning (ML) algorithms have led to the development of artificial intelligence (AI)-powered systems that can assist radiologists in detecting and diagnosing pneumonia. The use of ML algorithms for pneumonia discovery from x-ray images is a rapidly evolving research area that has the ability to improve the accuracy and efficiency of pneumonia diagnosis. ML algorithms can help overcome some of these challenges by analyzing large datasets of x-ray images to identify patterns and features that are indicative of pneumonia. By training on these datasets, these algorithms can learn to recognize the subtle differences between normal and pneumonia x-ray images, enabling them to make accurate diagnoses and assist healthcare professionals in making treatment decisions [5, 6]. Traditional ML algorithms have been widely applied to pneumonia detection on images of CXRs. ML is an interesting and potentially fruitful area of research that has the ability to improve the diagnostic accuracy of pneumonia as well as the efficiency with which it can be performed to apply ML algorithms to identify pneumonia from x-ray images.

BACKGROUND

Detecting pneumonia on CXR images has been the subject of a number of studies that have investigated the use of conventional ML techniques as well as the extraction of features using deep learning models.

Researchers employed the support vector machine (SVM) to diagnose pneumonia of COVID-19 patients from the x-ray images. This was done because SVM has a wonderful property that produces excellent results utilizing a tiny dataset. The image processing technique known as RGB and the SqueezeNet models were utilized to obtain additional photos to diagnose the dataset that was available. When applied to CXR pictures, our chosen model demonstrates an accuracy of 98.8% in identifying patients afflicted with COVID-19 [7]. To differentiate between normal x-rays and those of patients who showed signs of pneumonia that might have been caused by either a viral or bacterial source, 12 models that had already been pre-trained on ImageNet were adjusted and used in their research. Standard classification metrics were computed for each model like precision and F1 score, and the vast majority of the architectures were successful in achieving considerable performance, as shown by the fact that they were able to achieve up to 84.46% on average for the F1-score when differentiating between the four categories that were identified [8]. For this investigation, a dataset titled "Chest X-Ray Pneumonia" that had been curated and made accessible to the public was used. This dataset included 5856 anteroposterior (AP) CXRs that were separated into 1583 normal, 4273 viral, and bacterial pneumonia cases. The examination of convolutional neural network (CNN) on the various datasets demonstrated that its performance did not decrease from the first dataset to any of the five datasets that had varying degrees of noise. It was shown that the normal datasets had an accuracy of 90.2%, a specificity of 76.1%, and a sensitivity of 98.7% [9]. Model results from three different types of ML algorithms (k-nearest neighbors [KNN], SVM, and random forest) have been studied. In the study, researchers used two separate publicly accessible image archives. The best models developed for the initial dataset, which was comprised of pediatric CXRs, had an accuracy of 83.3% for radiomics, 89.9% for fractal dimension, and 91.3% overall. The second step included using data culled from a database of images made specifically for the purpose of studying COVID-19. The top models developed for this dataset achieved an accuracy of 95.3% in radiomics, 99% in fractal dimension, and 99% overall [10]. The RYOLO v4-tiny classifier was created by integrating the residual network with the YOLO extracting of features network. They developed a fully annotated dataset of non-COVID pneumonia CT scan and x-ray pictures to train and test the classifier. Computed tomographic (CT) scans provided these pictures. The suggested classifier has 11% higher F1 score, 10% higher accuracy, and 14% higher recall than the YOLO v4-tiny classifier. Medical professionals may use the RYOLO v4-tiny classifier to autonomously evaluate and identify COVID and non-COVID pneumonia from the CT scans and x-ray [11]. A hybrid social group optimization technique was used to find and choose suitable CXR image features. Various classifiers classified CXR images based on the qualities that were chosen. Support vector classifiers aid the pipeline in classification [12].

MATERIAL AND METHODS

Dataset and Preprocessing

The x-ray images that were used were obtained by participating in a competition that was held on Kaggle and was supplied by Paul Mooney. The contest consisted of a total of 5863 CXR images, including both patients with pneumonia and those who were otherwise healthy [4]. Throughout the process of our investigation, we gathered images intending to use 80% of them for training purposes and 20% for testing.

The first datasets brought to light several concerns that need further investigation. First, resize the image so that its dimensions match those of the original file, which is 224 × 224 pixels. The image that is used to train ML models must have the same dimensions as the raw images as these images come in a variety of sizes. Because of this, quick picture scaling is required to be ready for ML models [13].

Feature Extraction

Feature extraction was accomplished by using pre-trained CNNs, namely VGG16 and ResNet50; the results of this endeavor were then flattened. The collected features were employed as input by each respective algorithm used for ML.

VGG16 AND ResNet50

Feature extraction can be performed using the VGG16 and ResNet50 CNN architectures. VGG16 is a deep CNN architecture that is commonly used for image recognition tasks. There are a total of 13 convolutional layers, 5 max-pooling layers, and 3 fully linked layers included within the architecture [14]. ResNet50 is a deep CNN architecture that is commonly used for image recognition tasks. The architecture includes 50 convolutional layers and is known for its ability to learn very deep representations of images [15]. The following is a stepwise description of how feature extraction is performed by VGG16 and ResNet50 and how these features can be fed to ML for classification:

Loading of the pre-trained VGG16 and ResNet50 models: In the initial phase of the process, the framework will need to have the VGG16 and ResNet50 models, which are pre-trained, loaded into it.

Remove top layers: The next step is to remove the top layers of the pre-trained VGG16 and ResNet50 models, which are responsible for performing classification. This is because we are interested in using the lower-level features learned by the convolutional layers.

Extract features: The outcome of the final convolutional layer of the VGG16 and ResNet50 model is then fed into a feature extraction pipeline. This pipeline consists of flattening the output of the convolutional layer into a one-dimensional (1D) vector and feeding it into a feature extractor function, principal component analysis (PCA) and t-distributed stochastic neighbor embedding (t-SNE) are two examples of the dimensionality reduction strategies that this method employs to extract a smaller set of features that can be utilized by the ML algorithms [16].

With the VGG16 and ResNet50 CNN architectures, feature extraction is accomplished by removing the model's uppermost layers. The technique of feature extraction in ResNet50 is comparable to other CNN architectures; however, the introduction of skip connections enables the ML model to study from the input image more successfully by utilizing a feature extraction pipeline for extracting features from the outcome of the final convolutional layer and then feeding these features to various ML algorithms.

ML ALGORITHMS

ML is a subfield within the umbrella of AI that enables computers to learn. It involves the use of algorithms to train models based on large datasets to predict outcomes or make decisions. ML algorithms can be employed in an extensive range of applications, including computer vision, and prediction. They can help automate tasks, improve decision-making, and identify patterns that would be difficult for humans to detect. As with any type of algorithm, the success of an ML algorithm depends on the quality and quantity of the data used to train it. It also requires careful tuning of parameters and careful consideration of ethical issues such as bias and fairness.

The following conventional ML algorithms were used in the reviewed studies for pneumonia detection on CXRs.

SVM

Finding the hyperplane that most effectively divides data points into their respective categories is the key to its operation. The hyperplane with the largest margin (the distance between it and each classes nearest data points) is chosen. A kernel function is used by SVM to convert the data into a higher-dimensional feature space. This makes it possible to locate the hyperplane with greater ease. The optimal hyperplane is then found by solving a quadratic problem. The decision boundary of the SVM is defined as:

$$w^T x + b = 0 \tag{2.1}$$

where *w*, *x*, and *b* are the weight, input, and bias vectors, respectively. The SVM classifier predicts a new input as belonging to a particular category or class; this depends on which side of the decision boundary it falls. Equation represents the hyperplane that separates the data into different classes in the feature space defined by the kernel function [17]. To identify the ideal hyperplane that maximizes the margin between the classes is the objective of the SVM method, which can be expressed as a quadratic optimization problem. This task seeks to discover the best hyperplane.

Logistic Regression

An important ML approach for classification problems is known as logistic regression (LR). A logistic function is employed to describe the likelihood of an input belonging to a certain category or class in this way. The input is converted by the logistic function into a value that falls between 0 and 1, and this value may be understood as the likelihood of adhering to the positive class. The LR equation is given by:

$$p(y=1|x) = \frac{1}{\left(1 + \exp\left(-\left(w^T x + b\right)\right)\right)} \tag{2.2}$$

where $p(y = 1|x)$ is the probability of input x that belongs to positive class, w is the weight vector, b is the bias term, and exp is the exponential function. The weights are learned using maximum likelihood estimation by lessening the negative log-likelihood of the data [18].

KNN

KNN works by finding the k-nearest neighbors of a new input in the training dataset and predicting its class based on the most common class among its neighbors. The value of k is the hyperparameter, which necessitates it to be chosen based on the dataset and problem at hand. The KNN equation for classification can be written as:

$$y = argmax\left(\frac{1}{k}\sum_{i=0}^{k}(y_i)\right) \tag{2.3}$$

where y is the predicted class of input, k is the number of neighbors, and y_i is the category or class of the ith nearest neighbor [19].

Decision Tree

The operation of a decision tree involves the construction of a tree-like model of choices and the potential outcomes associated with those decisions. The tree's leaf nodes reflect class labels or regression values, whereas its inside nodes represent feature-based decisions. The equation for the decision tree can be written as:

$$f(x) = \left(\sum_{i=1}^{M}(c_i * I(x))\right) \tag{2.4}$$

where $f(x)$ is the predicted outcome for input x, c_i is the outcome for region R_i, and $I(x)$ where $x \in R_i$ is an indicator function that takes the value 1 if x belongs to the region R_i and 0 otherwise [20].

Random Forest

Random forest is an ML technique that works by first forming a huge number of trees of decisions and then merging the results of all of those trees to arrive at a single forecast. The training of each individual tree in the forest is accomplished by using a unique randomized subset of the data as well as a unique randomized subset of the characteristics. This helps to avoid the model from being overfit, which in turn enhances the performance of the model when it comes to generalization. The equation for random forest can be written as:

$$f(x) = \left(\frac{1}{B}\sum_{i=1}^{B}(f_i(x))\right) \tag{2.5}$$

where $f(x)$ is the predicted outcome for input x, $f_i(x)$ is the outcome of ith decision tree, and B is the total number of trees in the forest [21].

XGBoost

XGBoost (extreme gradient boosting) is a variant of gradient boosting that uses a more regularized model to prevent overfitting and improve performance. It works by constructing a series of decision trees, where each subsequent tree tries to spot the errors of the previous tree and corrects it. The XGBoost equation can be written as:

$$f(x) = \left(\sum_{i=1}^{k} (f_k(x)) \right) \tag{2.6}$$

where $f(x)$ is the predicted outcome for input x, k is the total number of trees, and $f_k(x)$ is the outcome of the kth decision tree [18].

AdaBoost

AdaBoost (adaptive boosting) works by merging numerous weak classifiers to create a powerful classifier. Each weak classifier is trained on a portion of the data, and its performance is then used to determine its weight. A weighted majority vote of the weak classifiers determines the final forecast. The AdaBoost equation can be written as:

$$f(x) = Sign\left(\sum_{i=1}^{M} (\alpha_i * h_i(x)) \right) \tag{2.7}$$

where $f(x)$ is the predicted outcome for input x, M is the total number of weak classifiers, $h_i(x)$ is the output of the ith weak classifier (which is typically a decision stump), and α_i is the weight assigned to the ith weak classifier. The weight α_i is calculated as:

$$\alpha_i = \left(\frac{1}{2} \ln((1 - \beta_i) / \beta_i) \right) \tag{2.8}$$

where β_i is the error rate of the ith weak classifier on the training data.

The algorithm iteratively trains new weak classifiers on the misclassified samples from the previous iteration and updates the weights of the samples to emphasize the misclassified samples [22]. This allows AdaBoost to focus on difficult samples and improve classification accuracy.

PERFORMANCE METRICS

Metrics employed to assess the effectiveness of ML models are known as performance metrics. Primarily, these metrics are used to assess classification models [image processing]. Accuracy, precision, sensitivity, specificity, recall, and F1 scores are examples. The analysis prioritized the accuracy and F1 scores above the others because they are the most often used and dependable [23].

RESULTS AND DISCUSSION

This investigation aimed to contrast the performance of several ML algorithms on the task of diagnosing pneumonia using the Paul Mooney dataset from Kaggle. The algorithms tested were SVM, LR, KNN, decision tree, random forest, XGBoost, and AdaBoost.

To conduct the research, the Paul Mooney dataset that was employed; it contains 5856 images of the CXR. The dataset was preprocessed and split into training, validation, and test sets, maintaining the balance between pneumonia and normal cases. Each image was labeled as either normal or pneumonia.

The images were preprocessed by resizing to 224 × 224 pixels and normalization. Image normalization is a technique used to standardize the pixel values of an image, ensuring that all images have the same scale and range of values. This is important for many computer vision tasks, as it can improve the performance and accuracy of ML algorithms.

The images were then input to the pre-trained VGG16 and ResNet50 for extracting relevant features for the ML algorithms. The extracted features were then used as input to each ML algorithm. Feature extraction is another important step that involves extracting relevant information or features from an image. Two popular models for feature extraction are VGG16 and ResNet50. VGG16 is a deep CNN that was trained on the ImageNet dataset and is known for its ability for extracting high-level features. ResNet50, on the other hand, uses residual connections to improve the training of the models, making it particularly effective for image recognition tasks. Both models can be used for feature extraction in a wide range of computer vision applications. The extracted features were further processed and standardized before being fed into the classifiers.

The findings of this study reveal that ML methods like XGBoost and AdaBoost are particularly effective for pneumonia diagnosis when using features extracted from ResNet50 CNN architecture. When compared with AdaBoost, which resulted in an accuracy of 93.10% and an F1 score of 0.917, the employment of XGBoost resulted in an accuracy of 94.40% and an F1 score of 0.926; this was the greatest accuracy that was achieved (Figure 2.2). Compared with the extracted features from VGG16, the ResNet50 extracted features performed well. The extracted features from VGG16 CNN architecture and results are presented in Figure 2.3. These algorithms' ability to combine the predictions from multiple weak learners may contribute to their superior performance, offering better generalization and robustness against overfitting.

However, the similar accuracy scores across all the algorithms suggest that the choice of ML algorithm may not be the most critical factor in achieving an accurate pneumonia diagnosis. Instead, the quality of the input features and the preprocessing steps may play a more significant role in determining the overall performance. From the results, it can be seen that all the algorithms achieved similar accuracy scores, with XGBoost, AdaBoost, and random forest performing the best with ResNet50

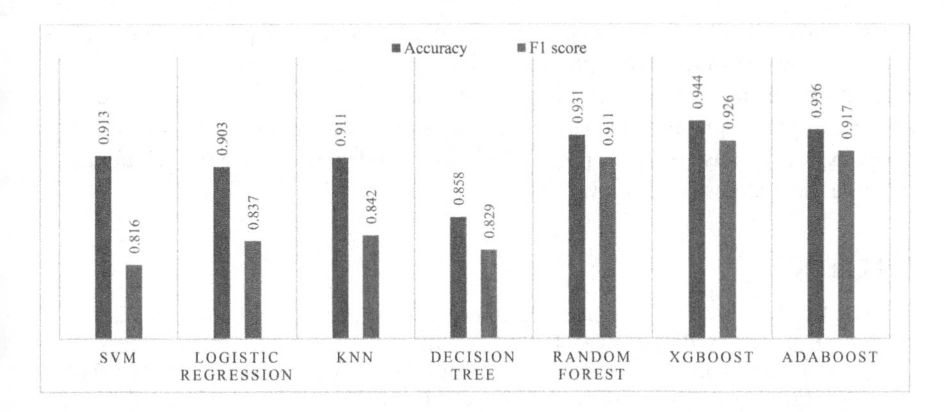

FIGURE 2.2 Shows the accuracy and F1 scores for each ML algorithm using extracted features from ResNet50 CNN architecture.

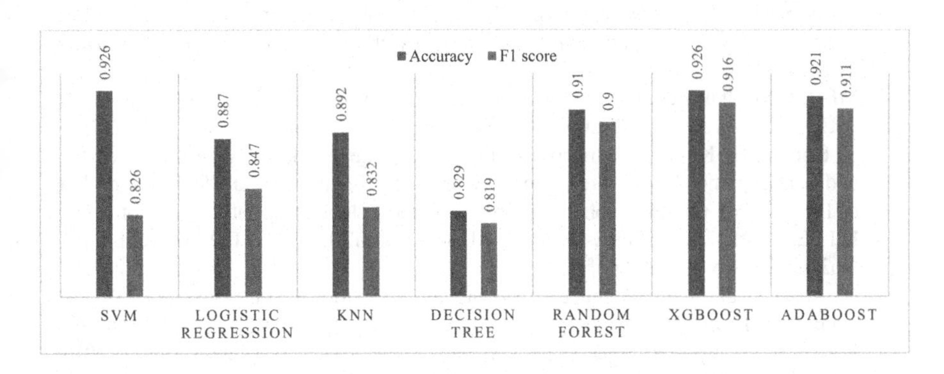

FIGURE 2.3 Shows the accuracy and F1 scores for each ML algorithm using extracted features from VGG16 CNN architecture.

features, whereas SVM performed the best with VGG16 features. These results indicate that using pre-trained CNN features for pneumonia diagnosis can achieve high accuracy regardless of the algorithm used. However, it is worth noting that some algorithms like decision tree performed relatively poorly, highlighting the importance of selecting appropriate algorithms for specific tasks.

CONCLUSION

The research provides valuable insights into the comparative performance of different ML algorithms for diagnosing pneumonia using features extracted from VGG16 and ResNet50. While XGBoost and AdaBoost outperform other methods, the choice of the algorithm may be less important than the quality of the input features and preprocessing steps. The use of XGBoost resulted in an accuracy of 94.40% and an F1 score of 0.926; this was the maximum accuracy that could be attained with ResNet50 extracted features. These findings can be used to guide the selection of algorithms for pneumonia diagnosis in clinical practice. ML pneumonia detection models should be tested using alternative feature extraction and preprocessing methods. With the continued development of these ML algorithms and advances in medical imaging technology, we can expect to see further improvements in the detection and treatment of pneumonia in the years to come. Furthermore, future research could explore the use of other pre-trained CNN models or other ML techniques to improve the diagnosis accuracy.

REFERENCES

1. G. Mackenzie, "The definition and classification of pneumonia," Pneumonia, vol. 8, no. 1, Aug. 2016, doi: 10.1186/s41479-016-0012-z.
2. WHO. "The top 10 causes of death," https://www.who.int/news-room/fact-sheets/detail/the-top-10-causes-of-death [Accessed: 01-Mar-2023].
3. DJ. Alapat, MV. Menon, Sh. Ashok, "A review on detection of pneumonia in chest X-ray images using neural networks," Journal of Biomedical Physics and Engineering, vol. 12, no. 6, Dec. 2022, doi: 10.31661/jbpe.v0i0.2202-1461.
4. P. Mooney, "Chest x-ray images (pneumonia)," Kaggle, 24-Mar-2018. [Online]. https://www.kaggle.com/datasets/paultimothymooney/chest-xray-pneumonia [Accessed: 10-Mar-2023].
5. MEH. Chowdhury et al. "Can AI help in screening viral and COVID-19 pneumonia?" IEEE Access, vol. 8, pp. 132665–132676, 2020, doi: 10.1109/access.2020.3010287.
6. S. Kumar, H. Kumar, "Lungcov: A diagnostic framework using machine learning and imaging modality," International Journal on Technical and Physical Problems of Engineering (IJTPE), vol. 14, no. 51, Number 2, June 2022. http://mail.iotpe.com/IJTPE/IJTPE-2022/IJTPE-Issue51-Vol14-No2-Jun2022/23-IJTPE-Issue51-Vol14-No2-Jun2022-pp190-199.pdf
7. N. Absar et al., "Development of a computer-aided tool for detection of COVID-19 pneumonia from CXR images using machine learning algorithm," Journal of Radiation Research and Applied Sciences, vol. 15, no. 1, pp. 32–43, Mar. 2022, doi: 10.1016/j.jrras.2022.02.002.
8. D. Avola, A. Bacciu, L. Cinque, A. Fagioli, M. R. Marini, R. Taiello, "Study on transfer learning capabilities for pneumonia classification in chest-x-rays images," Computer Methods and Programs in Biomedicine, vol. 221, p. 106833, Jun. 2022, doi: 10.1016/j.cmpb.2022.106833.
9. M. W. Kusk, S. Lysdahlgaard, "The effect of Gaussian noise on pneumonia detection on chest radiographs, using convolutional neural networks," Radiography, vol. 29, no. 1, pp. 38–43, Jan. 2023, doi: 10.1016/j.radi.2022.09.011.
10. C. Ortiz-Toro, A. García-Pedrero, M. Lillo-Saavedra, C. Gonzalo-Martín, "Automatic detection of pneumonia in chest X-ray images using textural features," Computers in Biology and Medicine, vol. 145, p. 105466, Jun. 2022, doi: 10.1016/j.compbiomed.2022.105466.
11. A. Kumar, "RYOLO v4-tiny: A deep learning based detector for detection of COVID and Non-COVID Pneumonia in CT scans and X-RAY images," Optik, vol. 268, p. 169786, Oct. 2022, doi: 10.1016/j.ijleo.2022.169786.
12. A. K. Singh, A. Kumar, M. Mahmud, M. S. Kaiser, A. Kishore, "COVID-19 infection detection from chest X-Ray images using hybrid social group optimization and support vector classifier," Cognitive Computation, Mar. 2021, doi: 10.1007/s12559-021-09848-3.
13. A. P. Ayshath Thabsheera, T. M. Thasleema, R. Rajesh, "Lung cancer detection using CT scan images: A review on various image processing techniques," Data Analytics and Learning, pp. 413–419, Nov. 2018, doi: 10.1007/978-981-13-2514-4_34.

14. K. Simonyan, A. Zisserman, "Very deep convolutional networks for large-scale image recognition," arXiv preprint, arXiv:1409.1556 preprint, pp. 1–14, 2014.
15. K. He, X. Zhang, S. Ren, J. Sun, Deep residual learning for image recognition, in: Proceedings of the IEEE Conference on Computer Vision and Pattern Recognition (CVPR), 2016, pp. 770–778, doi: 10.1109/CVPR.2016.90
16. S. Sharma, K. Guleria, S. Tiwari, S. Kumar, "A deep learning based convolutional neural network model with VGG16 feature extractor for the detection of Alzheimer disease using MRI scans," Measurement: Sensors, vol. 24, p. 100506, Dec. 2022, doi: 10.1016/j.measen.2022.100506.
17. L. P. Fávero, P. Belfiore, R. de Freitas Souza, "Support vector machines," Data Science, Analytics and Machine Learning with R, pp. 323–370, 2023, doi: 10.1016/b978-0-12-824271-1.00032-9.
18. C. Dong et al., "Non-contact screening system based for COVID-19 on XGBoost and logistic regression," Computers in Biology and Medicine, vol. 141, p. 105003, Feb. 2022, doi: 10.1016/j.compbiomed.2021.105003.
19. H. Arslan and H. Arslan, "A new COVID-19 detection method from human genome sequences using CpG island features and KNN classifier," Engineering Science and Technology, an International Journal, vol. 24, no. 4, pp. 839–847, Aug. 2021, doi: 10.1016/j.jestch.2020.12.026.
20. A. A. Dehghani, N. Movahedi, K. Ghorbani, S. Eslamian, "Decision tree algorithms," Handbook of Hydroinformatics, pp. 171–187, 2023, doi: 10.1016/b978-0-12-821285-1.00004-x.
21. M. Hemalatha, "A hybrid random forest deep learning classifier empowered edge cloud architecture for COVID-19 and pneumonia detection," Expert Systems with Applications, vol. 210, p. 118227, Dec. 2022, doi: 10.1016/j.eswa.2022.118227.
22. R. Wang, "AdaBoost for feature selection, classification and its relation with SVM, a review," Physics Procedia, vol. 25, pp. 800–807, 2012, doi: 10.1016/j.phpro.2012.03.160.
23. S. Kumar, H. Kumar, R. Agarwal, V. K. Pathak, "Human disease prognosis and diagnosis using machine learning," Lecture Notes in Electrical Engineering, pp. 41–53, 2022, doi: 10.1007/978-981-19-0284-0_4.

3 Enrichment of Human Life through Intelligent Wearable Technology

Santhiya S, Jayadharshini P, Abinaya N,
Sharmila C, Vasugi M, and
Swetha Nallamangai K N
Kongu Engineering College, Perundurai, Tamil Nadu

INTRODUCTION

As there is a rapid proliferation in the information and communication technology, there is an increase in number of new services and user demand. One such service that has emerged from this demand is a "wearable device" or a "wearable technology" or simply "wearables" [1]. These terms refer to small electronic and mobile devices that can be incorporated into gadgets, jewels, accessories, or clothes. These wearables are designed to be worn on a user's body. The most sophisticated examples of wearables are the artificial intelligence(AI) hearing aid, Google glasses, and virtual reality (VR) [2]. The advancements in digital technology have increased the development of new devices, and one such developed device is a wearable device. Wearable technology includes any device that is designed to be worn on the user's body, controlled by software that is linked to the device. Recently the need for and popularity of wearable devices has increased. In the near future wearable devices could become an indispensable part of our life. Wearables are be released into the universe slowly and steadily. With the help of various modern technologies, wearable devices are more sophisticated to satisfy the needs of everyone [3]. All the hardware and software components of this devices are interconnected to provide the expected work to be done as expected.

There are three textile fundamental classes of wearables:

i. Wearable well-being technology: this includes the devices used in the healthcare sectors.
ii. Wearable technology: this field includes the making of clothes and other textiles.
iii. Wearable consumer technology: this class includes examples like a wristwatch, a wristband, rings, chains, etc.

Even though there are many pros, cons exist that may cause harm to human life. Thus Figure 3.1 gives an overview of the various types of wearable devices.

DOI: 10.1201/9781003388319-3

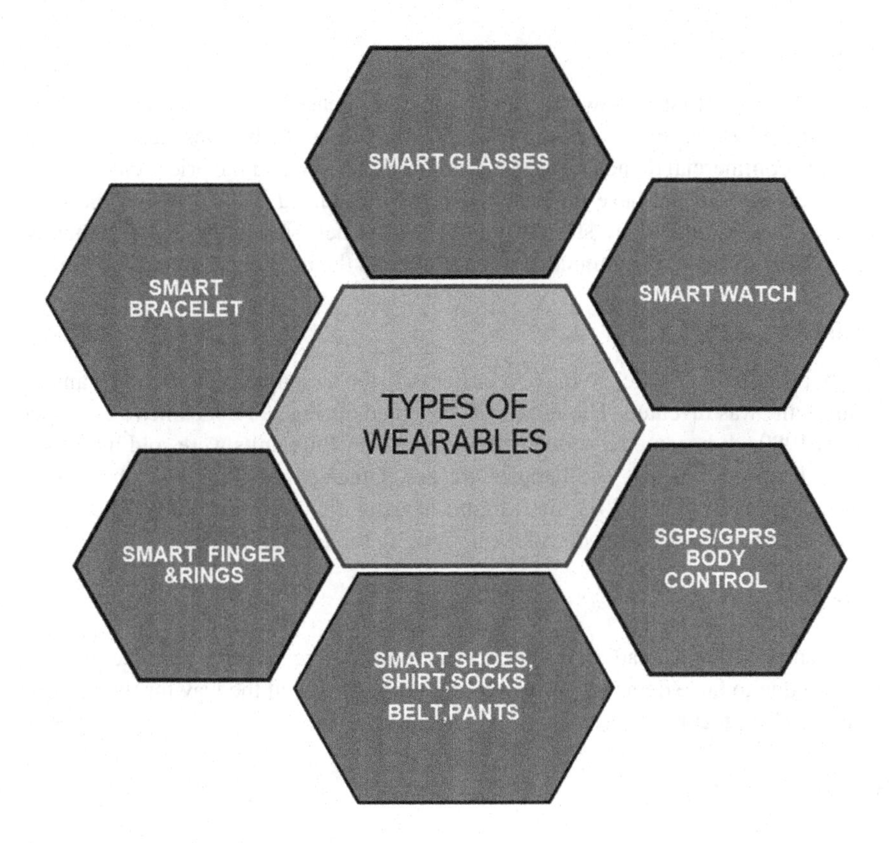

FIGURE 3.1 Types of wearables.

HISTORY

Wearable technology started its new-fangled spin in the 1950s. Since the 1950s it has undergone several transitions with several new technologies incorporated into various new devices.

1950s

Radio

Wearable technology began with Sony's transistor radio. It was the first transistor radio designed by Sony and premiered in 1955. This transistor radio, also known as TR-55, serves as a template for today's portable gadgets.

Virtual Reality

Sometimes the most progressive technology gets invented before a simpler one, for example, VR. In the 1950s, Morton Helig worked on creating the Telesphere Mask, the first head-mounted display that rendered film in 3D and stereo sound to the user.

1970s

During 1975 the first wristwatch calculator was released. It became iconic in the 1970s and 1980s. In the beginning it was costly, but as the business leaders understood the commercial demand, the technologies improved and the prices came down.

Technology did not have much forward growth during the 1970s, but the technologies that we use today started then. The first head-mounted camera emerged during 1970s. The head-mounted camera stand was the beginning for today's GoPro.

1980s

We are all found of music. To make it easier to listen to songs anywhere we want the "Walkman" was invented. Though it was designed during 1979, it came to existence only in 1980s. It became so popular that over 200 million units were sold initially.

The healthier we are, the happier we are. During the 1980s wearable health devices emerged. In, 1987 the first digital hearing aids were available. The fitness tracker was also born in this decade, emerging in the form of heart monitors.

1990s

In December 1994, Canadian researcher Steve Mann created the wearable wireless webcam due to huge demand. This wearable webcam paved the way for the Internet of things (IoT) technologies.

2000s

As discussed earlier, the Walkman emerged during the 1980s and moved a step forward taking the form of an iPod. Thousands of songs could be saved and played whenever the listener wanted. It was portable and much easier to carry to gyms and while walking.

Bluetooth, a source of communication between electronic devices, also emerged in this decade. It has had a great impact and we are using wireless Bluetooth today for talking over phones.

The most used health tracker, Fitbit, emerged during late 2009. The founder of Fitbit hoped for about 50 pre-orders during its release, but there were only 5 pre-orders. As its popularity increased over 25,000 Fitbits were sold.

2010s and Beyond

The wearable craze greatly increased during this period of time. Google introduced the wearable glasses, and Apple introduced its first iWatch. Today most of us have at least one wearable device be it a health tracker or accessories.

APPLICATION OF WEARABLE DEVICES

Healthcare

One of the most crucial aspects of life is health. Health is wealth, as they say! Without pretty good health, you really can't do much, thus it is important to look after oneself. In the chaos of regular living, it is simple to overlook this. Fortunately, there are numerous forms of wearable healthcare technology that can aid in keeping that in mind [4]. There are, for instance, posture aids that can assist in maintaining proper posture to avoid back pain. The matchbox-sized Upright Go posture trainer is an illustration of this. When you slump forward in your posture, it will gently vibrate when you attach it to your upper back.

Technology has advanced over the past several decades to the point where wearable device systems can now continually collect data in real time and integrate on-body sensors for monitoring physiological signals without discomfort. In this regard, wearable devices hold the promise of assisting in the initial diagnosis as well as ongoing care for a large number of people with neurological, cardiovascular, and pulmonary illnesses. Since the invention of transistors, wearable devices have shrunk significantly, and the IoT has made them smarter, more portable, multipurpose, and able to connect to the Internet while also operating on batteries. Numerous applications, such as those in communication, sports, optics, VR, and AI, have made use of these devices [5].

Fitness

Since its inception, wearable technology has been mostly utilized in the fitness industry. Yes, I'm referring to the Fitbit, which has become so well liked that it is now more of a fashion accessory than a fitness tracker. For those who are unaware, Fitbit is a fitness company that develops smart watches and activity monitoring bands that keep track of fitness-related metrics. Fitbit devices are linked to a fitness app that displays information about the wearer's daily activity, sleep objectives, etc. Furthermore, the user may link more than one Fitbit to the app, and it will instantly sync with everything they are wearing. Figure 3.2 clearly visualizes various fitness trackers used today [6].

Another wearable technology is synchronized with clothing. For example, yoga pants can be Bluetooth-synchronized, and until you shift into the proper yoga postures, they will lightly electrically pulse your hips, knees, and ankles. On your phone, you can view additional comments regarding your yoga postures. That is one method of practicing yoga without a teacher! The Google commuter jacket, which was created in collaboration with Levi's, is another example of smart apparel. It's a brilliant bicycling jacket that enables touch-free navigation and audio control without requiring you to touch your phone. This jacket is innovative in a variety of ways!

Jewelry

Who doesn't adore jewelry, whether it be lovely necklaces or adorable rings? It is stylish, amusing, and takes your costume to the next level. However, with the advancement of wearable electronics, jewelry has many more functions besides just being attractive. Smart rings are available that can keep an eye on our health and send all the required info to our smartphone. The Oura Ring, a smart ring that can measure physical health and keep an eye on our vitals while we sleep, is one example of this. What if you're not a fan of rings? Fear not! You can match any of your earrings with earring backings made by Joule. These monitor your heart rate, degree of

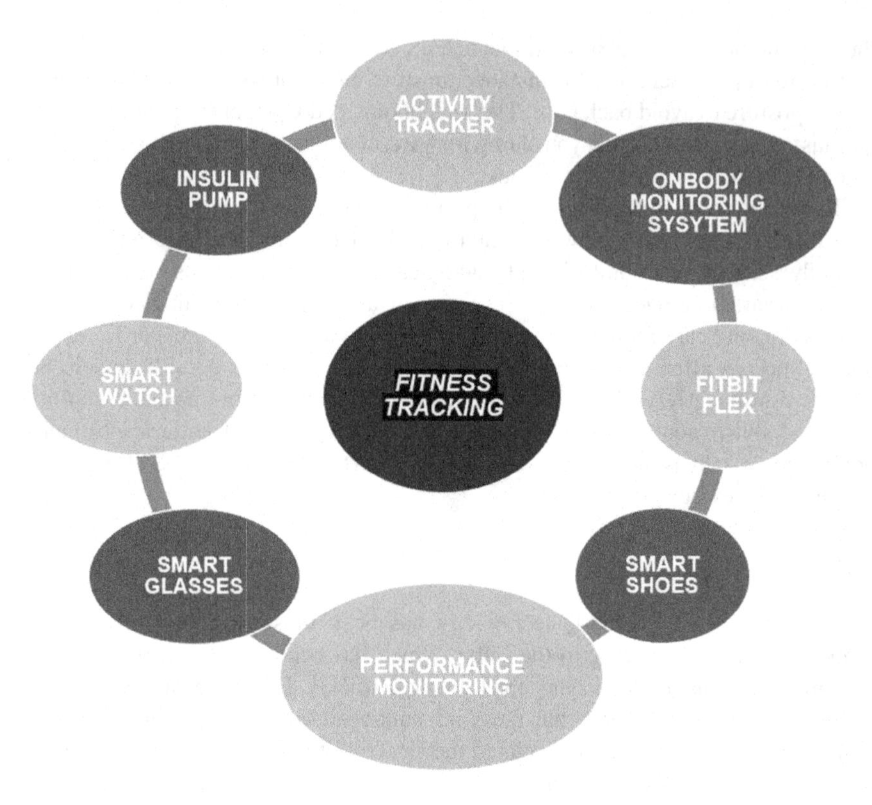

FIGURE 3.2 Fitness tracking.

exercise, and number of calories burned. And because it has GPS, you don't have to worry about your smart jewelry being stolen. You may now look stylish and maintain your health!

Security

Have you seen those spy flicks where the protagonist is equipped with glasses that double as tiny computers that can scan the antagonist and provide all the information? Yes, thanks to a combination of wearable technology, computer vision, and facial recognition, those glasses might actually come to pass. Consider the possibility that law enforcement officers could be seen sporting what appear to be regular eyeglasses that can be used to identify criminals by facially matching their image to one in the police database. It would be even better if security cameras were connected to these smart glasses via the IoT so that police officers could know if there were any suspects nearby.

Child Security

Child safety wearables are a result of children's growing need for safety. This wearable's main advantage is that it is compatible with any cell phone and doesn't need to be operated by a tech-savvy person. The majority of wearable devices on the market today are designed to send parents information on their children's whereabouts, activities, etc., via Wi-Fi and Bluetooth, which seem to be unreliable communication channels.

This tool is designed to make it simple for parents to find their children and to rapidly notify them when those children are in dangerous situations like fire or gas leaks. As a result, this chapter aims to provide a GSM mobile communication-enabled SMS text-enabled communication medium between the parent and the child's wearable. This wearable will then respond with a text stating the child's accurate locations [7].

The child's parent can send texts with specific keywords such as "location," "temperature," etc. The wearable device will then reply back with a text containing the necessary details about the child. The device warns those nearby so they may take immediate action to keep the youngster safe until the parents arrive or they can get in touch with the parents and ask for assistance in finding them.

In this case, the wearable device's SOS light feature would be activated if a parent received an SMS alerting them that their child was lost and might be found. As the SOS signal is widely recognized as the signal required for help, this signal will alert those nearby about the child when the child is in need of assistance.

An ATmega328p microcontroller with an Arduino Uno boot loader serves as the foundation and controller for the wearable's system architecture [8]. The GSM shield acts as an interface to deliver data that are received by SMS or MMS to a smartphone. The Arduino Uno uses the GSM shield as a trigger to ask its various modules for data.

The Arduino GSM shield causes the Arduino Uno to request the most recent GPS coordinates if a text message with distinct characters asking for the user's current location or GPS coordinates are transmitted to the Arduino GSM shield via the user's smartphone. The GSM shield then sends the coordinates to the user's smartphone via SMS after the Arduino Uno processes the coordinate information it has received and transferred to it.

MARKET OVERVIEW OF WEARABLES

The rise in usage of AI assistants and integration of wireless technology has increased the market demand for wearables.

Figure 3.3 clearly shows the market demand for AI-driven wearable devices. There has been an intense growth in the use of the wearables. As the usage increases, the demand increases accordingly. It also shows the forecast of the market demands for 2024. It is significantly high when compared with other years. The global wearable AI market size was valued at $10.63 billion in 2019 and is projected to reach $69.51 billion by 2026, and to register a compound annual growth rate (CAGR) of 26.50% during the forecast period [9].

The global wearable market is divided into segments on the basis of the products, technology, application, and geography. This segmentation analysis of various countries, by application and by products, is shown in Figure 3.4.

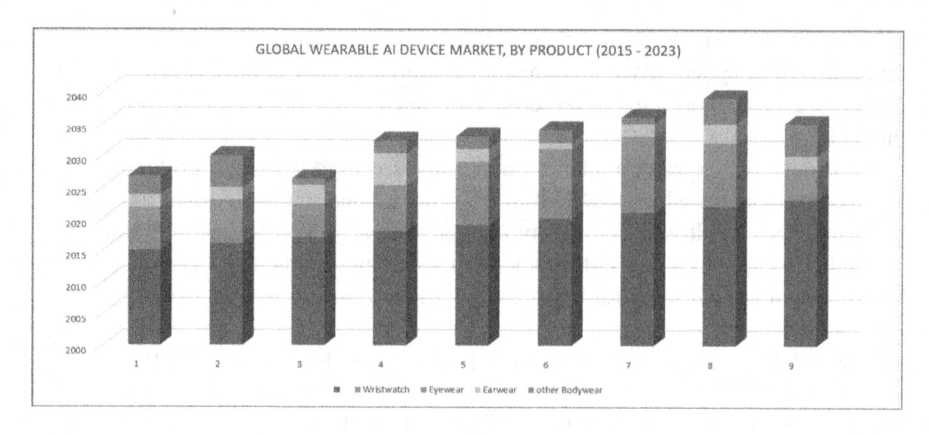

FIGURE 3.3 Forecast of market growth.

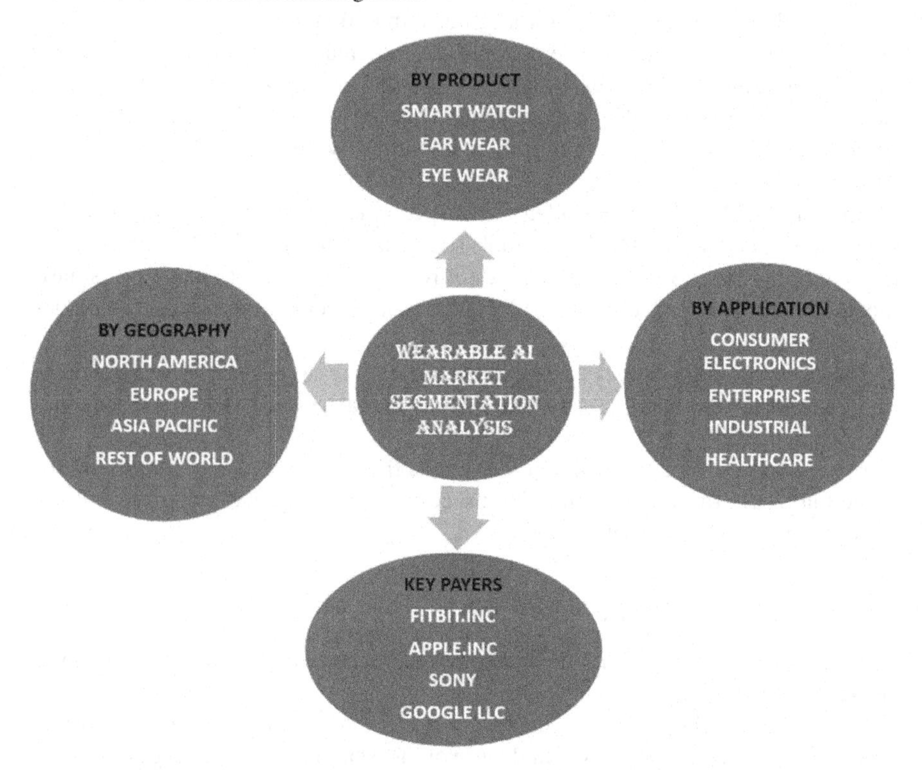

FIGURE 3.4 Segmentation analysis.

Wearable Market by Product

These products are classified into smart watch, ear wear, eye wear and others [6].

Figure 3.3 shows the market analysis of the wearable products. Here the smart watches hold the lion's share of the products, with an obvious increase from 2000 to 2023.

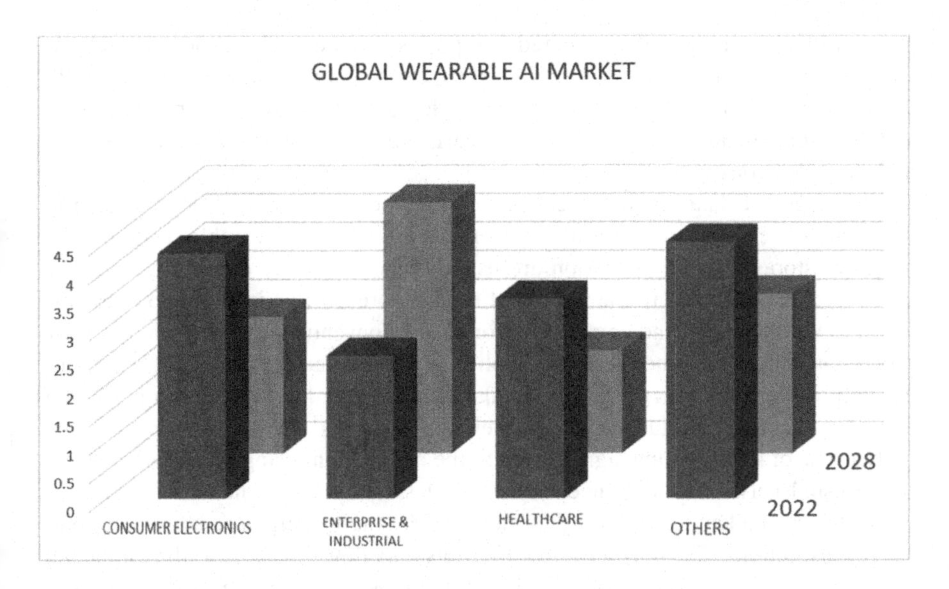

FIGURE 3.5 Market analysis by application.

Wearable Market by Technology

This segment includes Wi-Fi, Bluetooth, GPS, NFS, ANT+ and others. The ANT+ segment is estimated to have the highest CAGR during the forecast period.

Wearable Market by Application

This segment includes consumer electronics, enterprise and industrial, healthcare, and others.

From Figure 3.5 it is obvious that consumer electronics hold the largest market for the forecasted period.

From all the previous analyses, it is clearly evident that market demand for wearables is increasing [10].

FRUITFUL WORK DONE

The following describe work done on wearable devices.

Microprocessor chips are becoming wearable due to their ever-increasing computational power and drop in cost. In other words, they can be incorporated into objects like clothing, jewelry, and even human beings. A new computing paradigm is emerging as a result of the development of wearable technologies. Wearable gadgets are attached to our bodies, expanding the limitations of our bodies with various capabilities such as sensing, communicating, and computing in contrast to traditional computers, which are machines separated from their users.

Although wearable technology comes in a variety of sizes, designs, and capacities, its primary purpose is typically collection of data and processing. They are basically little versions of connected computers. As wearable devices are physically attached to individuals, the data they collect about their health and GPS whereabouts often have a strong privacy element. Activity trackers and smart watches, for instance, may track users' heart rates, exact steps, sleeping patterns, and even locations.

The stages of data sensing, local processing, local storage, and transmission to a remote server or storage are all part of the data flow in wearable devices. The stages of storage and transmission are usually where security or privacy breaches take place. Even so, due to the variety of hardware resources, including processors, RAM, power supplies, and communication range, conventional security and privacy protection solutions cannot be easily applied to every wearable devices. The battery capacity of wearable devices, in particular, represents a key obstacle to providing extensive functionality due to size and weight constraints.

Because of its important significance to the health monitoring system, the design of a trusted network is now needed for wireless body sensor networks. Wireless networks are the key to unlocking 24-hour patient monitoring both within and outside of institutions. The telemetric link permits data flow from and to the implanted sensor device in both directions. The implanted node's tiny memory unit contains a record of the sensed data. The wireless programming facility is offered by the over-the-air programming facility due to the telemetric linkage between the sink and the implanted node [11].

PROS OF WEARABLES

- Because of wearable technology, people can check their fitness levels, track their whereabouts via GPS, and more.
- They can view text messages faster.
- The best part of wearables is that the majority of the gadgets that let us accomplish such things are hands-free and portable, so we don't need to take them out of our pockets.
- These wearables are connected to our smart devices, which sends the information to them and allows us to access it both now and in the future.
- We can use this to set goals and monitor our progress toward them.
- Multitasking is easier.
- Wearables help to schedule our day.
- They are more comfortable [12].

CONS OF WEARABLES

Every technology has its own pros and cons. In such a way the cons of the wearables are discussed next.

- Source of distraction: Because the wearables are connected to a device like mobile phones, the notifications might cause a distraction for the user. For example, let us consider a smart watch wearable connected to a mobile phone. In this case the user decides to put away the mobile phone and concentrate on work, but when a notification is received in the mobile phone the watch lights up or vibrates, creating a distraction. This has been the greatest source of distraction for drivers and desk employees.

- Expensive: Wearables are also available at a cheaper rate. But if we need more specifications and benefits the cost increases accordingly. So, it can be heavy on the wallet [13].
- Battery life: Most wearable devices come with a battery limitation. The Fitbits have a battery life for about 2–3 days. An Apple iWatch only lasts for a day or so. In the hectic work schedule in this busy world, we might forget to charge these devices, and they discharge all too quickly.
- Health hazards: Most wearables are either connected through Wi-Fi or Bluetooth. The information is transmitted via radio waves, which may in turn cause health hazards [14].

CONCLUSION

From this analysis of wearable technologies, we can infer that with this technology one can track their fitness and use it for security purposes as well. So, with the help of this technology and in the future, wearables will be more useful will become one of the most popular modern technologies in the world.

REFERENCES

1. Bonato, P. Wearable Sensors/Systems and Their Impact on Biomedical Engineering. IEEE Eng Med Biol Mag 2003, 22, 18–20.
2. Philips.com. 'Philips Design SKIN Probe Receives Prestigious "best of the best" in Red Dot Award: Design Concept 2007. N.p., 2014. Web. 20 May 2014.
3. Williamson, J.; Liu, Q.; Lu, F.; Mohrman, W.; Li, K.; Dick, R.; Shang, L. Data Sensing and Analysis: Challenges for Wearables. In Proceedings of the20th Asia and South Pacific Design Automation Conference, Chiba, Japan, 19–22 January 2015; pp. 136–141.
4. Altizer, R., Jr; José, P.; Zagal, Johnson, E.; Bob, W. Design Box Case Study: Facilitating Interdisciplinary Col-Laboration and Participatory Design in Game Develops. In Extended Abstracts Publication of the Annual Symposium on Computer-Human Interaction in Play, 2017; pp. 405–412.
5. Nugroho, J. A Conceptual Framework for Designing Wearable Technology. University of Technology Sydney, 2013.
6. Mardonova, M.; Choi, Y. Review of Wearable Device technology and Its Applications to the mining industry. Energies 2018, 11(3). https://doi.org/10.3390/en11030547
7. Robinson, R.; Isbister, K.; Rubin, Z. All the Feels: Introducing Biometric Data to Online Gameplay Streams. In CHI PLAY'16 Extended Abstracts, 2016; pp. 261–267. https://doi.org/10.1145/2968120.2987732
8. Rich, E.; Miah, A. Mobile, Wearable and Ingestible Health Technologies: Towards a Critical Research Agenda. Health Sociol Rev 2017, 26, 84–97. [Google Scholar]
9. E G See, Stacey Higginbotham Cloud-Platform-for Tracking-Your-Health, Archived at https://perma.cc/AX5L-M87W; see Also Samsung Unwraps Tizen for 'Internet of Things Samsung Launches a Wearable Wristband and Cloud Platform for Tracking Your Health, GIGAOM, Volume 11 Posted: 2014-05-28.

10. Pearce, M. Growing Use of Police Body Cameras Raises Privacy Concerns. Los Angeles Times. Los Angeles Times, 27 Sept. 2014. Web. 04 Mar. 2017.
11. Mavrogiorgou, A.; Kiourtis, A.; Kyriazis, D. A Pluggable IoT Middleware for Integrating Data of Wearable Medical Devices. Smart Health 2022, 26, 100326. [Google Scholar]
12. GCF Global. Pros and Cons of Wearable Technology. https://edu.gcfglobal.org/en/wearables/pros-and-cons-of-wearable-technology/1/
13. Kahraman, B.; Vakhter, V.; Costanzo, I.; Bu, G.; Foroozan, F.; Guler, U. A Miniaturized Prototype for Continuous Noninvasive Transcutaneous Oxygen Monitoring. In 2022 IEEE Biomedical Circuits and Systems Conference (BioCAS), 2022; pp. 486–490.
14. Vowles, C. J.; Van Engelen, S. N.; Noyek, S. E.; Fayed, N.; Davies, T. C. The Use of Conductive Lycra Fabric in the Prototype Design of a Wearable Device to Monitor Physiological Signals. In 2022 44th Annual International Conference of the IEEE Engineering in Medicine & Biology Society (EMBC), 2022; pp. 922–925.

4 Reliability and Validity of Survey Questionnaires for Identifying Learning Disabilities in an Intelligent Tutoring System

Neelu Jyothi Ahuja
University of Petroleum and Energy Studies, Dehradun, Uttarakhand

Sarthika Dutt and Swati Arya
COER University, Roorkee, Uttarakhand

INTRODUCTION

A learning difficulty is a neurological condition categorized by difficulties in reading, writing, calculation, and anger management that results in in a child's lack of education (Kohli et al., 2018). It is one of the conditions that need to be diagnosed early for successful child development. Dyslexia, dysgraphia, and dyscalculia are learning disabilities that have been linked to the inability to read, illegible handwriting, spelling mistakes, and inability to perform well with arithmetic operators. Either these conditions could impact the learner's cognitive skills alone or present as a comorbid issue with other difficulties. In India psychologists are the at the frontline for detecting and providing therapies to the learning disabled. These therapists manage common difficulties related to learning-disabled learners Robinson (2009); however, only a handful of psychologists are experts in the field (Alborz et al., 2005). Several studies indicated the gaps in knowledge in solving learning difficulties issues in developing countries like India; addressing these issues is required to implement effective strategies to manage learning difficulties. The 19-item questionnaire designed into an intelligent tutoring system (ITS) for identifying learning disabilities was tested for its reliability and validity. This questionnaire was primarily developed with the aim of identifying learning disabilities among children in the state of Uttarakhand, India (Dutt et al., 2021). Then this questionnaire was validated and tested in several sample sizes. Therefore, the objective of this study was to adapt and validate this questionnaire among the learning-disabled learners in local settings.

DOI: 10.1201/9781003388319-4

In the present study survey questionnaire reliability and validity have been analyzed to determine the acceptability of the questions involved in the pretest. This tool can be utilized effectively to identify the learning disabilities of the learning-disabled learners. Learning disability identification is a complex process and requires multiple screenings if the tools are not reliable and validated. Therefore, this study is significant as it can aid special educators, psychologists, and parents in identifying a child's problematic skills. The present study is organized as follows. The Literature Review section explores the studies and research carried out in past to test the reliability and validity of questionnaires. In the Methodology section, the survey questionnaire to identify learning disabilities was tested. The Results and Discussion section reviews the results obtained in this study followed by conclusion and future scope in the last section.

LITERATURE REVIEW

The preliminary work in a research study is to identify respondents before providing them assistive support. For that purpose, the researcher needs to conduct a survey based on empirical research on certain parameters like planning of questionnaire, preparation of questionnaire design, and its validation Taherdoost (2016b). A great deal of research work is available for conducting the survey. Some of the important related research studies are discussed next.

In a book by Warwick & Lininger (1975), the author shared their experience with a training program held at the Employment and Human Resources Service (SERH) of the ministry Labor of Peru. Rossi et al. (2013) performed a questionnaire survey with the help of word processing software instead of telephonic questionnaire, which at that time was a novel approach. This book also discussed the performance of a mixed-mode questionnaire, i.e., via email, telephone, Internet, etc. They concluded that mixed (multi)-mode surveys were found to be the most popular. Raghunathan & Grizzle (1995) used a questionnaire survey to reduce the financial burden of recruiting and selecting individuals for studies. They shared their motivation to use and experience with a questionnaire survey of the Cancer Prevention Research Program. The target sample population was male or female, 18 years old, and from the state of Washington. The survey was carried out by telephonic conversation where various questions were asked regarding the prevalence of risk behavior regarding cancer such as alcohol consumption, smoking, dietary habits, etc. Because the questionnaire time was about 50 minutes instead of 30 minutes, there was a large non-response rate. The result of the questionnaire survey was that 70% of the later questions didn't get a response, whereas in 26% of the cases family members refused to respond. Ballinger & Davey (1998) designed and used a questionnaire to avoid some of these potential risks. They tried to eliminate the misconception that sometimes the reader thought that no special preparation or training was required to produce a questionnaire and that it was relatively easy. Bee & Murdoch-Eaton (2016) tried to help the reader create good questionnaire design in three simple steps: preparation, evaluation, and delivery. Analysis and interpretation were the end stages of accomplishing the research work. The emphasis was on wording because inappropriate wording could result in jargon, which should be reduced to avoid potential confusion. Closed questions can easily handle a large amount of data. The response scale was also used. This scale takes input from 0 to 10 for scaling and provides the output in the form of emojis. Authors are advised to include researchers, stakeholders, and respondents to evaluate their questionnaire, and to make acceptability and feasibility important factors for consideration. Oosterveld et al. (2019) introduced a new method of testing goals called the taxonomy linking procedure for guiding test developers and users. Taxonomy made the assumption that construction techniques are focused on psychometric elements. It distinguishes between six various techniques. The logical approach makes use of expert opinions to guarantee face validity. Prototypicality assessments are used in the prototypical technique to guarantee process validity Aithal et al. (2015). In the internal procedure, homogeneity-enhancing item sets are chosen. By choosing items that most accurately predict an external criterion, the external method maximizes criterion validity de Jong et al. (2018). The construct method maximizes construct validity by considering theoretical factors. Due to comprehensive representation of the concept domain, the facet technique aims to maximize content validity. Kachroo & Kachen (2018) presented a methodology that emphasized linear cases only for the placement of the items of surveys. Since then, they have been focused on obtaining optimal reliability so they developed a mathematical model that was based on a number of specific structured assumptions to test the reliability and consistency. They transformed the general problem into a mathematical optimization problem Aithal & Aithal (2019). Validity measures question relevancy and reliability, therefore, a questionnaire's consistency (internal) should be measured through several metrics.

Sirakaya-Turk & Uysal (2017) covered important domains for data collection and generation. One domain was underlying logic and principles of survey research and means of data collection. Further, domains include issues in designing survey instruments. In the applied research field, sampling and design of survey research is one of the important aspects of data collection techniques. The authors discussed the logic of sampling and representativeness, choosing representative units of analysis, sampling process and methods, the concept and terminology of sampling, sample size, confidence intervals, and sample error (variability) Choy (2014). After that, they addressed various issues, including how to develop and design a questionnaire, how to develop effective questions in a questionnaire, how to evaluate questionnaires, and how to conduct surveys and secondary analysis of collected survey data. Finally, the authors concluded their work by briefly considering special issues included in Internet and panel surveys by quoting their pros and cons in different data collection methods. Moroney & Joyce (2019) motivated the audience by saying anyone can write a questionnaire by asking the right questions to the right person at the right time while using an appropriate data collection process to retrieve the required information. They provided their guidance for preparing an appropriate questionnaire. Yaddanapudi & Yaddanapudi (2019) elaborated on the guidelines to prepare a questionnaire with respect to three survey-based studies. They emphasized that surveys found to be economical were efficient in collecting data, especially in cases of a large population, in a short period of time.

Various machine learning tools have been used such as the binary classifier using supervised machine learning, support vector machine (SVM), and random forest for (RF) the detection of developmental dyslexia using eye movement Aithal (2016b). SVM was used as a baseline method so that comparisons could be made easier and clearer. In this work, SVM mapped its input vectors into a high dimensional feature space via chosen nonlinear mapping and then found an optimal hyperplane to separate the classes with a maximal margin, which reduces generalization error Aithal (2016a). Whereas RF used as a classifier comprising an ensemble of randomized decision trees, which made a mutual decision on the class, as we did with SVM (Raatikainen et al., 2021). In one work more than 3600 participants were chosen as trained data, and the proposed model showed that 80% of the participants were correctly detected with dyslexia. A new dataset of 1300 participants with age customization were tested in a different environment. The researchers provided a tablet instead of a desktop computer and found a recall of over 78% for the class with dyslexic children 12 years old or older. This work showed that dyslexia can be screened using a machine learning approach. An online screening tool in Spanish based on this method has already been used (Rello et al., 2019). This study has been conducted to analyze the post high-school outcomes of exited high-school students who were suffering from intellectual disability and were on the autism spectrum from a southwestern U.S. state Kothari (2004). Authors have used a predictive analytics approach to analyze these students' outcomes data, in a regular interval under U.S. special-education law. Data modeling was conducted using machine learning and logistic regression Huck (2007). The strongest and significant predictors in the findings were the students who were spending at least 80% of their instructional days in general education settings and those who were graduating from high school (Yamamoto & Alverson, 2022).

Bovell-Benjamin et al. (2009) and Riaño & Palomino (2015) discussed survey studies of dietary practices. Peytchev & Peytcheva (2017) performed a survey to reduce the measurement error that occurs due to survey length. They proposed a design approach for the evaluation of the split questionnaire. According to the authors, respondents may find lengthy survey instruments demanding, which could lead to increased measurement error. The association between length and measurement error has minimal empirical support, which might cause surveys to go on for longer than is ideal. Finding ways to cut survey length while still achieving survey goals is at least as important as creating the survey itself. In this study, a split questionnaire design (in which respondents were randomly assigned to receive different subsets of the survey modules) is used to investigate the feasibility of reducing measurement error associated with survey length Oluwatayo (2012). The missing information is then multiplicatively imputed for each respondent. However, any advantages to survey estimates from the lengthening of the survey may be outweighed by the imputation variation. To further assess the impact, we employ an experimental approach Onwuegbuzie et al. (2009).

METHODOLOGY

Participants

The participants included in this study were the learners from government and private schools along with children from Doon Psychotherapeutic Center in Dehradun in the State of Uttarakhand, India.

Materials

A pretest questionnaire was developed in March 2020 in India, UK (Uttarakhand).

Procedure

This study was carried out using the learner (with and without) learning disabilities. This study was conducted in two parts. In the first part the questionnaire is validated, and in second part the learner's feedback on the validated questionnaire is analyzed. This work presents the detailed method and analyses of the validation of questions and feedback evaluation. This is a cross-sectional study that has observed the questionnaire reliability and validation that has been considered in the research for identification of learning disabilities. The reliability testing includes internal consistency analysis. Figure 4.1 depicts the overall process flow diagram of this study. First pretest questionnaire was developed for children in the age-group 5–12 years. This questionnaire framework includes information regarding learning disabilities (dyslexia, dysgraphia, and dyscalculia), type/subtype, associated cognitive skills, learning styles, and problematic skills of the learners. The content was then piloted and its validity analyzed. Then the analysis of the validity and internal consistency of the items in this study on 94 learning-disabled or non-learning-disabled learners using exploratory factor analysis to test construct validity and Cronbach alpha test to identify the internal consistency was done.

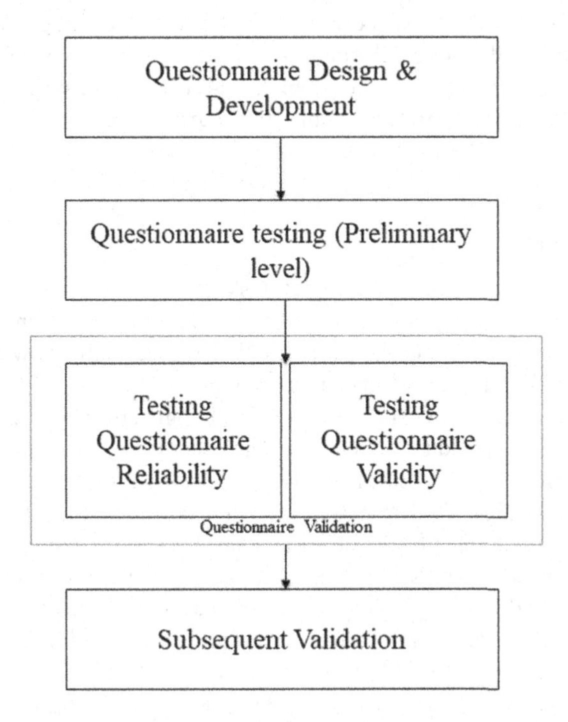

FIGURE 4.1 Process flow diagram.

Design

The questionnaire draft is completed through initial pilot testing and successive revisions from the experts; the initial validation of this framework is among the participants done through psychometric evaluation. The evaluation involved psychometric construct validity and reliability testing of the developed content. The construct validity was performed by analyzing the significant difference between the two groups to establish known groups validity. The construct validity was performed by analyzing the significant difference between the two groups to establish validity of the known group (Rohrbach et al., 2022). The hypothesis formulated was that non-learning-disabled learners would have better knowledge regarding the reading, writing, and basic math skills compared with learning-disabled learners. Reliability testing involved internal consistency testing of the questionnaire framework using Cronbach's reliability test (Bujang et al., 2018). Test-retest reliability is analyzed using Cohen's kappa statistics to measure the stability of the responses to the questionnaire (Tan et al., 2015). The questionnaire is in simple naive language and comprises a total of 38 items categorized into two grades, i.e., grade 1 and grade 2, based on the level of the learner's knowledge. There were seven items covering the following areas in dyslexia identification: (1) phonological awareness, (2) random naming, (3) fluency, (4) loudness while reading, (5) non-sense words, (6) literacy, and (7) cognitive strength. For the dysgraphia identification it comprises seven items indicating the following difficulties: (1) visual spatial response, (2) motor skills, (3) sentence structure, (4) word formation, (5) jumbled words, (6) space knowledge, and (7) spellings. The other five items were used to indicate the difficulties of learners with dyscalculia: (1) reasoning, (2) short-term memory, (3) word problems, (4) basic math skills, and (5) shapes identification. these questionnaire items and learners' responses were recorded using the Likert scale (0–5 scale).

Content Validation

Validating the questionnaire is a complex task. It is the process of analyzing the dependability of the survey questions. They are analyzed by identifying what is intended to be measured using the survey tool. It also explains how accurately the data are collected in the area of experimentation.

In the first stage, a team of experts of one psychologist, two special educators, and one trainer were introduced to the questionnaire plan for learning-disabled learners and were also asked to plan for structuring the design for questionnaires to identify learning disabilities. They reviewed the items considered for the questionnaire framework. The guidelines for the development of assistive technologies and tools for learning-disabled learners was also reviewed by experts (Hayes et al., 2018). The experts verified the questionnaire as per the current guidelines and recommendations including recent innovations and support for learning-disabled learners (Darling-Hammond et al., 2020). The sufficient number of items to represent the questionnaire framework were also analyzed under this process. **Questionnaire framework:** Then the questionnaires were arranged into two grades as per the level

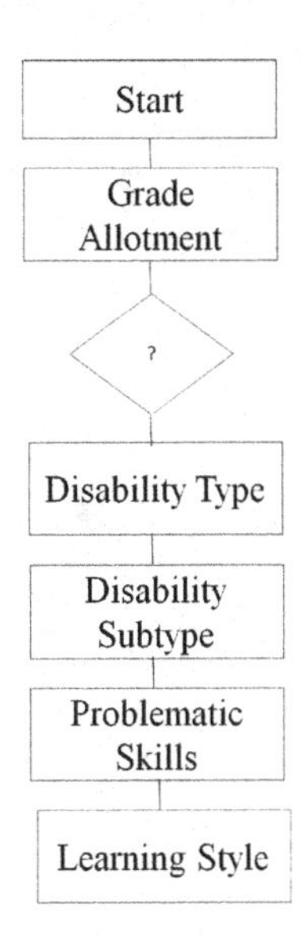

FIGURE 4.2 Conceptual framework.

of the learner's knowledge. The questionnaire was aligned to determine the learners' learning difficulties using type/subtype (dyslexia, dysgraphia, and dyscalculia), cognitive strength, learning style, and problematic skills of the learners (Dutt et al., 2021). All the questions were in English. Figure 4.2 shows the conceptual framework of the designed questionnaire framework and the associated details with considered disabilities (dyslexia, dysgraphia, and dyscalculia). The conceptual framework of type of learning disability, cognitive strength, learning style, and problematic skills about the learning disabilities of the learners is represented in Figure 4.2. The framework is translated into an online computer-based test and incorporated within a learner model of an ITS depicted in Figure 4.3. Consent was given by the parents/teachers or special educators prior to the experiment conducted with the participants. The information regarding the pretest questionnaire and the consent form were available on the developed system. The questionnaire could be accessed after submitting the consent form. The learners first register themselves or log into the system, and then they are assigned to a particular grade questionnaire, either grade 1 or grade 2, based on their level of knowledge. The level of difficulty increases from grade 1 to grade 2.

Face Validation

The questionnaire was self-administered and the learner's feedback on the same was recorded. The feedback questions include their concepts and understanding regarding language, content, and the learner module interface while interacting with the system. The corrections and suggestions based on the learner's feedback are incorporated into the developed framework represented in Figure 4.3.

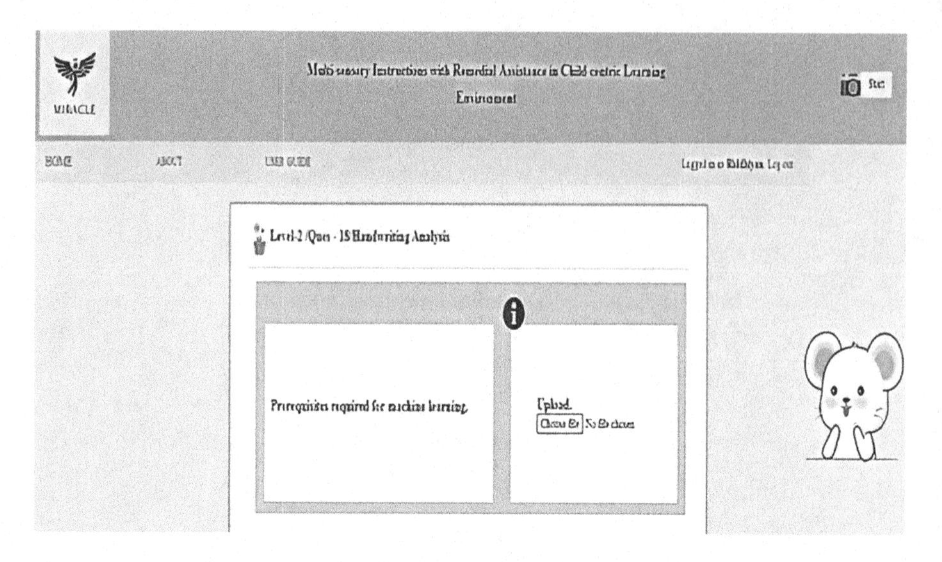

FIGURE 4.3 Intelligent tutoring system platform.

RESULTS AND DISCUSSION

CHARACTERISTICS OF SURVEY QUESTIONNAIRES

The number of questionnaire items was categorized into two grades (grade 1 and grade 2) with respect to the learner's level of knowledge and age. Compliance was high (nearly 80%) for the pretest of the learning-disabled learners, with almost all learners completing the questionnaire. Seventy-eight learners completed the questionnaire at least once (62 non-learning-disabled and 16 learning-disabled learners). There was a significant gender and age difference in the study population ($p = 0.004$). Table 4.1 represents the gender and age of the participants.

Table 4.2 presents the learners who achieve higher than the other group learners in both grades of the questionnaire ($p < 0.005$). The learners who were non-learning disabled had an average score of (Mean ± SD)9.4 ± 5.7 and learning-disabled learners

TABLE 4.1
Gender and Age of the Participants

<table>
<tr><th></th><th>Learning-Disabled Learners (n = 30)</th><th>Non-Learning-Disabled Learners (64)</th></tr>
<tr><td>Age range of learning-disabled learners</td><td>8–12 years</td><td>7–12 years</td></tr>
<tr><td>Boys</td><td>10</td><td>46</td></tr>
<tr><td>Girls</td><td>9</td><td>29</td></tr>
<tr><td>Special educators/teachers</td><td colspan="2">32</td></tr>
<tr><td>Total population</td><td>19</td><td>75</td></tr>
</table>

TABLE 4.2
Scores of Learners in Mean, Standard Deviation, and Range (n = 94)

Grades	Learning-Disabled Learners (n = 30)				Non-Learning-Disabled Learners (64)			
	Minimum	Maximum	Mean	SD	Minimum	Maximum	Mean	SD
Grade 1 (19 items)	0	19	2.4	1.4	1	19	5.5	0.8
Grade 2 (19 items)	0	18	5.0	2.9	0	18	5.1	0.9
Total (38)	0	26	7.6	3.4	2	27	11.5	3.9

had an average score of 7.4 ± 4.7 out of a maximum score of 19 in the pretest questionnaire. The learners with learning disabilities and non-learning disabilities performed equally well in posttest assessments: non-learning disabled with an average score of 10.8 ± 7.7 and learning disabled with 9.9 ± 8.7.

The gender difference in the two groups was balanced and controlled for the analysis of covariance, but it did not affect the overall scores and results. Also, analysis of covariance on age did not affect the significant results generated during analysis.

Internal Consistency Reliability of Survey Questionnaire

In the case of missing values regression, imputation is selected to replace these values. Item means ranged from 3.00 to 6.26 on the 5-point Likert scale and standard deviation ranged from 0.53 to 1.56, showings some variability. There were low to moderate significant correlations present among items from the dyslexia (DYS), dysgraphia (DYG), and dyscalculia (DYC) subscales. The questionnaire's subscale reliability was accessed using Cronbach's alpha optimal values (0.7–0.9). ENG (Learner's Engagement), WTU (Willingness to Use System) here demonstrated as "very good" to "excellent" values for Cronbach's alpha and no items were removed from these subscales. The Cronbach's alpha value for the USE subscale (0.694) was acceptable, so removing the item "I will discontinue to use the ITS interface" has improved the consistency in the subscales. The means obtained for ENG, WTU and USE subscales were appropriate for the 5-point Likert scale (toward the midpoint) as given in Table 4.3.

TABLE 4.3
Scales Used in the Current Study

Grades	Internal Reliability (Alpha Value)		
	Non-Learning Disabled	Learning Disabled	Overall
Grade 1 (19 items)	0.670	0.703	0.79
Grade 2 (19 items)	0.562	0.783	0.79
Total (38)	0.702	0.852	0.89

The Cronbach's alpha values for learning-disabled and non-learning-disabled learners obtained were significantly different (please refer to Table 4.3). The values of Cronbach's alpha obtained (when considered individually) for learning-disabled learners were higher than the non-learning-disabled learners. When all learners were considered, the Cronbach's alpha value obtained was 0.79 for both grades. The alpha value for the complete questionnaire with two grades was 0.89.

The significant difference between the score of the learners (with or without) learning difficulties indicated that the survey questionnaire had satisfactory construct validity. The non-learning-disabled learners scored higher in both grades. These normal learners gave more correct responses in pretest questionnaires. However, the learners (with or without) learning difficulties have performed equally well in post-test questionnaires. This implies that learning-disabled learners have experienced a positive impact from pretest to posttest analysis.

CONCLUSION

The survey questionnaire accessed in the present study for identifying learning disabilities integrated with an artificial intelligence-based ITS proved to be a valid and reliable assisting tool. The survey questionnaires were simple and easy to interpret by the learners. They can be utilized as a reliable tool to determine the learner's cognitive strength, learning style, type/subtype of disability and its severity, and problematic skills of the learners. When implementing a questionnaire to identify learning disabilities, ITS could be used to explore remote areas where people are not even aware of learning disabilities. This information could be profitably utilized for awareness and to improve learning-disabled conditions in society.

REFERENCES

Aithal, P. S. (2016a). Review on Various Ideal System Models Used to Improve the Characteristics of Practical Systems. International Journal of Applied and Advanced Scientific Research, 1(1), 47–56. http://doi.org/10.5281/zenodo.159749

Aithal, P. S. (2016b). Study on ABCD Analysis Technique for Business Models, Business Strategies, Operating Concepts & Business Systems. International Journal in Management and Social Science, 4(1), 98–115. http://doi.org/10.5281/zenodo.161137

Aithal, P. S., & Aithal, S. (2019). New Directions in Scholarly Research- Some Fearless Innovations & Predictions for 21st Century Research. International Journal of Management, Technology, and Social Sciences (IJMTS), 4(1), 1–19. https://doi.org/10.5281/zenodo.2557222

Aithal, P. S., Shailashree, V. T., & Suresh Kumar, P. M. (2015). A New ABCD Technique to Analyze Business Models & Concepts. International Journal of Management, IT and Engineering (IJMIE), 5(4), 409–423. http://doi.org/10.5281/zenodo.61652

Alborz, A., McNally, R., & Glendinning, C. (2005). Access to Health Care for People With Learning Disabilities in the UK: Mapping the Issues and Reviewing the Evidence. Journal of Health Services Research & Policy, 10(3), 173–182. https://doi.org/10.1258/1355819054338997

Ballinger, C., & Davey, C. (1998). Designing a Questionnaire: An Overview. British Journal of Occupational Therapy, 61(12), 547–550.

Bee, D. T., & Murdoch-Eaton, D. (2016). Questionnaire Design: The Good, the Bad and the Pitfalls. Archives of Disease in Childhood-Education and Practice, 101(4), 210–212.

Bovell-Benjamin, A. C., Dawkin, N., Pace, R. D., & Shikany, J. M. (2009). Use of Focus Groups to Understand African-Americans' Dietary Practices: Implications for Modifying a Food Frequency Questionnaire. Preventive Medicine, 48(6), 549–554.

Bujang, M. A., Omar, E. D., & Baharum, N. A. (2018). A Review on Sample Size Determination for Cronbach's Alpha Test: A Simple Guide for Researchers. The Malaysian Journal of Medical Sciences: MJMS, 25(6), 85–99. https://doi.org/10.21315/mjms2018.25.6.9

Carmines, E. G., & Zeller, R. A. (1979). Reliability and Validity Assessment, Newbury Park, CA, SAGE.

Choy, L. T. (2014). The Strengths and Weaknesses of Research Methodology: Comparison and Complimentary between Qualitative and Quantitative Approaches. IOSR Journal of Humanities and Social Science, 19(4), 99–104.

Darling-Hammond, L., Flook, L., Cook-Harvey, C., Barron, B., & Osher, D. (2020). Implications for Educational Practice of the Science of Learning and Development. Applied Developmental Science, 24(2), 97–140. https://doi.org/10.1080/10888691.2018.1537791

de Jong, J. A., Dorer, B., Lee, S., Yan, T., & Villar, A. (2018). Overview of Questionnaire Design and Testing. Advances in Comparative Survey Methods: Multinational, Multiregional, and Multicultural Contexts (3MC), Somerset, NJ, John Accept change Wiley & Sons, 115.

Dutt, S., Ahuja, N. J., & Kumar, M. (2021). An Intelligent Tutoring System Architecture Based on Fuzzy Neural Network (FNN) for Special Education of Learning Disabled Learners. Education and Information Technologies, 1–21. https://doi.org/10.1007/s10639-021-10713-x

Hayes, A. M., Dombrowski, E., & Shefcyk, A., et al. (2018 Apr). Learning Disabilities Screening and Evaluation Guide for Low- and Middle-Income Countries [Internet], Research Triangle Park (NC), RTI Press.

Huck, S. W. (2007). Reading Statistics and Research, Needham Heights, MA, Allyn & Bacon.

Kachroo, P., & Kachen, S. (2018). Item Placement for Questionnaire Design for Optimal Reliability. Journal of Marketing Analytics, 6(4), 120–126.

Kohli, A., Sharma, S., & Padhy, S. K. (2018). Specific Learning Disabilities: Issues That Remain Unanswered. Indian Journal of Psychological Medicine, 40(5), 399–405. https://doi.org/10.4103/IJPSYM.IJPSYM_86_18

Kothari, C. R. (2004). Research Methodology: Methods and Techniques, New Delhi, New Age International.

Lewis, S. (2015). Qualitative Inquiry and Research Design: Choosing Among Five Approaches. Health Promotion Practice, 16(4), 473–475.

Moroney, W. F., & Joyce, A. (2019). Questionnaire Design: How to Ask the Right Questions of the Right People at the Right Time to Get the Information You Need. User's Guides to Human Factors and Ergonomics Methods. Washington, DC, Human Factors and Ergonomics Society.

Oluwatayo, J. A. (2012). Validity and Reliability Issues in Educational Research. Journal of Educational and Social Research, 2(2), 391–400.

Onwuegbuzie, A. J., Dickinson, W. B., Leech, N. L., & Zoran, A. G. (2009). A Qualitative Framework for Collecting and Analyzing Data in Focus Group Research. International Journal of Qualitative Methods, 8(3), 1–21.

Oosterveld, P., Vorst, H. C., & Smits, N. (2019). Methods for Questionnaire Design: A Taxonomy Linking Procedures to Test Goals. Quality of Life Research, 28(9), 2501–2512.

Peytchev, A., & Peytcheva, E. (2017, December). Reduction of Measurement Error Due to Survey Length: Evaluation of the Split Questionnaire Design Approach. Survey Research Methods, 11(4), 361–368.

Raatikainen, P., Hautala, J., Loberg, O., Kärkkäinen, T., Leppänen, P., & Nieminen, P. (2021). Detection of Developmental Dyslexia With Machine Learning Using Eye Movement Data. Array, 12, Article 100087. https://doi.org/10.1016/j.array.2021.100087

Raghunathan, T. E., & Grizzle, J. E. (1995). A split Questionnaire Survey Design. Journal of the American Statistical Association, 90(429), 54–63.

Rello, L., Baeza-Yates, R., Ali, A., Serra, M., Bigham, J. P., & Protopapas, A. (2019). Predicting Risk of Dyslexia With an Online Gamified Test. PLoS One, 15(12), e0241687. https://doi.org/10.1371/journal.pone.0241687

Riaño, C. E., & Palomino, M. (2015). Design and Preparation of a Questionnaire According to Delphi Method to Select Virtual Laboratories (VL). Sophia, 11(2), 129–141.

Robinson, W. S. (2009). Ecological Correlations and the Behavior of Individuals. International Journal of Epidemiology, 38(2), 337–341.

Rohrbach, P. J., Dingemans, A. E., & Essers, B. A. et al. (2022). The ICECAP-A Instrument for Capabilities: Assessment of Construct Validity and Test–Retest Reliability in a General Dutch Population. Quality of Life Research, 31, 687–696. https://doi.org/10.1007/s11136-021-02980-5

Rossi, P. H., Wright, J. D., & Anderson, A. B. (Eds.). (2013). Handbook of Survey Research. Burlington, MA, Elsevier Science.

Sirakaya-Turk, E., & Uysal, M. S. (2017). Survey Research: Sampling and Questionnaire Design. Research Methods for Leisure, Recreation and Tourism, (2 ed.), 122–146.

Taherdoost, H. (2016b). Validity and Reliability of the Research Instrument; How to Test the Validation of a Questionnaire/Survey in a Research. International Journal of Academic Research in Management (IJARM), 5(3), 28–36.

Tan, C. L., Hassali, M. A., Saleem, F., Shafie, A. A., Aljadhey, H., & Gan, V. B. (2015). Development, Test-Retest Reliability and Validity of the Pharmacy Value-Added Services Questionnaire (PVASQ). Pharmacy Practice, 13(3), 598. https://doi.org/10.18549/PharmPract.2015.03.598

Warwick, D. P., & Lininger, C. A. (1975). The sample survey: Theory and practice. McGraw-Hill.

Yaddanapudi, S., & Yaddanapudi, L. N. (2019). How to Design a Questionnaire. Indian Journal of Anaesthesia, 63(5), 335.

Yamamoto, S. H., & Alverson, C. Y. (2022). Post-High School Outcomes of Students with Autism Spectrum Disorder and Students with Intellectual Disability: Utilizing Predictive Analytics and State Data for Decision Making. Journal of Intellectual Disabilities, https://doi.org/10.1177/17446295221100039

5 A Survey of Artificial Intelligent Techniques for Cancer Detection

Neha Nandal
Gokaraju Rangaraju Institute of Engineering and Technology, Hyderabad, Telangana, India

Rohit Tanwar
University of Petroleum and Energy Studies, Dehradun, Uttarakhand, India

INTRODUCTION

Artificial Intelligence (AI) is very much in demand due to recent technology developments for detection of cancer. AI-based technologies, such as deep learning algorithms, image analysis, and natural language processing (NLP), are being used in a variety of ways to improve the accuracy and efficiency of cancer diagnosis. One area where AI has been particularly useful is in the analysis of medical images like magnetic resonance imaging (MRI), computed tomography (CT) scans, and mammograms. AI-based algorithms can be trained to identify abnormalities in these images, such as tumors, which can be an early indicator of cancer. This can help with early diagnosis and treatment, which can improve patient outcomes.

Another area where AI has been applied is in the analysis of patient data like electronic health records (EHRs), lab results, and genetic data. AI-based algorithms can analyze these data to identify patterns and predict outcomes, such as the likelihood of developing cancer or the response to treatment. This can help with personalized medicine and precision oncology. Several studies have shown that AI-based technologies can improve the accuracy of cancer detection compared with traditional methods. For example, a study published in The *Lancet Digital Health* found that a deep learning algorithm was able to accurately identify lung cancer on CT scans with an Appropriate Use Criteria (AUC) of 0.96, which is significantly better than the AUC of 0.76 for radiologists. However, there are also challenges with using AI in cancer detection. For example, there is a need for large amounts of data to train these algorithms, and there are concerns about bias in the data. Additionally, there are concerns about the interpretability of these models, and the potential for over diagnosis. Overall, AI in cancer detection is a promising field that has an efficiency of cancer diagnosis, but there are also challenges that must be addressed.

Cancer refers to the diseases that occur due to spreading of abnormal cell and unbalanced growth that is uncontrollable. Different types of cancer have their own

DOI: 10.1201/9781003388319-5

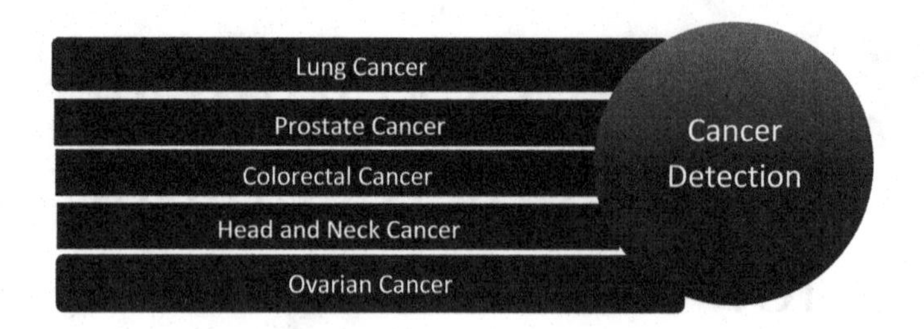

FIGURE 5.1 Different types of cancer detection using artificial intelligence.

symptoms, causes, and cures. Some common types of cancer include lung, breast, prostate, and colorectal. Each type of cancer develops in a specific part of the body and has its own set of characteristic signs and symptoms. Different symptoms for the same can be unexplained pain, skin changes, bladder or bowel movements, or weight loss. Cancer can also spread to different parts of the body, a process called metastasis.

Several factors can be the reason to increase the risk of cancer, such as bad lifestyle, exposure to harmful chemicals, environmental factors, and genetics. The risk can be handled with changing lifestyle, quitting smoking, having a healthy routine, and sun protection. The treatment for cancer can vary depending on the type and stage of the cancer, but it often includes surgery, chemotherapy, and radiation treatment. Newer treatments such as immunotherapy, targeted therapy, and precision medicine are being developed to improve the treatment and the prognosis of cancer patients. The different ways to utilize AI for cancer detection are shown in Figure 5.1.

There are many types of cancer that can be detected using AI, including the following:

- Lung cancer: AI-based algorithms can analyze CT scans to identify abnormalities in the lungs, such as tumors, which can be an early indicator of lung cancer.
- Breast cancer: AI-based algorithms can analyze mammograms to identify abnormalities in the breast tissue, such as masses or calcifications, which can be an early indicator of breast cancer.
- Prostate cancer: AI-based algorithms can analyze MRI images to identify abnormalities in the prostate, such as tumors, which can be an early indicator of prostate cancer.
- Skin cancer: AI-based algorithms can analyze images of skin lesions, such as moles, to identify abnormalities that may indicate skin cancer.
- Colorectal cancer: AI-based algorithms can analyze images of the colon and rectum, such as colonoscopy images, to identify abnormalities that may indicate colorectal cancer.
- Head and neck Cancer: AI-based algorithms can analyze images of head and neck, such as CT or MRI images, to identify abnormalities that may indicate head and neck cancer.
- Blood cancer (leukemia): AI-based algorithms can analyze blood samples to identify biomarkers that may indicate the presence of blood cancer, such as abnormal white blood cells.

- Ovarian cancer: AI-based algorithms can analyze images of the ovaries and pelvic area, such as ultrasound images, to identify abnormalities that may indicate ovarian cancer.

These are some examples of how AI can be used in cancer detection, but it is important to note that AI models are not perfect and they need to be validated through clinical trials and independent studies before they can be used in a clinical setting. Additionally, AI-based cancer detection is a rapidly growing field, new technology and research are emerging to improve the accuracy and efficiency of cancer diagnoses.

REVIEW PLAN

The review of multiple studies to provide a more accurate estimate of the overall effect of a particular treatment or intervention has been done. The strategy for the review is shown in Figure 5.2. Conduct of a comprehensive search for relevant studies was made, which includes searching multiple databases, such as PubMed and Embase, and manually searching reference lists and contacting experts in the field. After identification of relevant research, the next step is to identify those studies that meet the criteria of inclusion for the analysis.

Once the studies have been selected, the next step is data extraction from the studies, which includes extracting data on the population, intervention, and outcome, as well as information on the study design, sample size, and results. The last step is the interpretation of the results, including the summary of the findings and the implications for practice or future research.

RELATED WORK

There are many examples of AI that people use in their daily lives [1]. Many smartphones now come with virtual assistants, such as Siri or Google Assistant, which utilize NLP to enable a response to voice. AI-based models are utilized by many online retailers for product recommendation on the basis of their purchase history. AI is used to curate content on social media platforms, such as Facebook and Instagram, and to personalize the news feed for each user. AI-based algorithms are used to recommend music and videos to users based on their listening and viewing history. Many

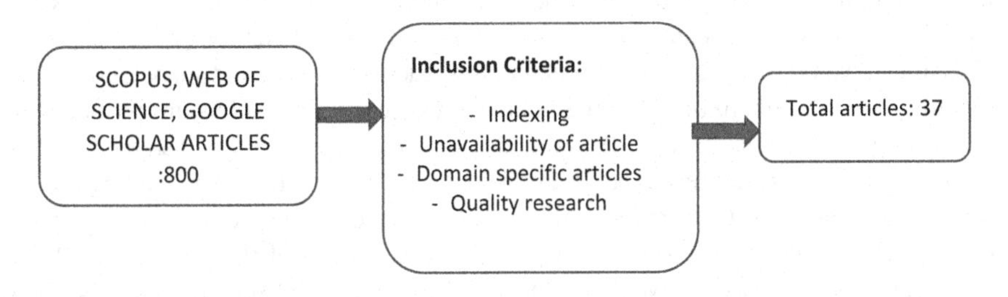

FIGURE 5.2 Review strategy.

navigation apps, such as Google Maps and Waze, use AI-based algorithms for providing updates on real-time traffic and for suggestion of fastest route. Many email clients use AI-based algorithms to filter out spam and to prioritize important messages. AI-based algorithms are used by financial institutions to detect and prevent fraudulent transactions. AI is used in healthcare for things like drug discovery, patient monitoring, and diagnostic imaging [2]. The association rule-based system for breast cancer detection and prevention uses association rule mining to identify relationships between different factors, such as age, family history, and lifestyle habits, and the risk of breast cancer. These relationships can then be used to develop personalized risk assessments and recommendations for breast cancer screening, prevention, and treatment.

In the context of predicting the types of human T-cell lymphotropic virus (HTLV), a hybrid machine learning model could involve combining multiple models such as neural networks, support vector machine (SVM), and decision trees for analyzing data like patient demographics, laboratory results, and genetic information [3]. One example of a hybrid machine learning model that has been used to predict the type of HTLV is a combination of a decision tree and a neural network. The decision tree is used for identification of relevant features in data, whereas the neural network is used to make the final prediction. The combination of these two models can improve the accuracy of predictions by leveraging the strengths of both models.

There are a few potential reasons why a person may experience nausea and dizziness after eating [4]. Knowledge-based collaborative deep learning (KCDL) is a method that uses deep learning techniques and expert knowledge to classify benign and malignant lung nodules on chest CT scans [5]. The goal of KCDL is to improve the accuracy and reliability of lung nodule classification by combining the strengths of both deep learning and human expertise.

One of the key benefits of hybrid convolutional neural networks (HCNNs) for breast tumor segmentation is their ability to handle large amounts of data and complex image features [6]. The algorithm can automatically learn to identify the location, shape, and size of the tumor, which is important for radiogenomics studies. Additionally, the hierarchical architecture allows for more accurate segmentation of breast tumors, which can improve the accuracy of radiogenomic analyses. Approaches like nearest feature line (NFL) and the adaptive neuro-fuzzy inference system (ANFIS) classifier typically involve preprocessing the medical images to extract relevant features, such as shape, size, and texture. These features are then used to train the NFL and ANFIS classifier, which can be used to classify new images based on their features [7]. The combination of NFL and the ANFIS classifier has been shown to improve the accuracy of medical image classification compared with traditional methods. An artificial neural network (ANN) can be used to create a malignancy risk prediction model for thyroid nodules. This model would use a variety of input features, such as the size and shape of the nodule, to predict the likelihood that the nodule is cancerous [8]. ANNs are particularly well suited for this task because they are able to learn complex relationships between input features and the target output (malignancy risk) through training on a large dataset of thyroid nodules. Once trained, the model can be used to predict the malignancy risk of new nodules and aid in the diagnosis and treatment of thyroid cancer [9]. A healthcare system based on wearable Internet of Things (IoT) sensors could be used to identify and control the spread of the chikungunya virus. The system would consist of wearable sensors that continuously monitor the patient's vital signs, including heart rate and body temperature, as well as other indicators of the virus, such as joint pain and rash. These sensor readings would be transmitted to a central server via a low-power wireless connection and analyzed using machine learning algorithms. mHealth, or mobile health [10], has the potential to greatly benefit aging populations. mHealth can make it easier for older adults in China to access healthcare services, such as telemedicine consultations and remote monitoring of chronic conditions. Joint weakly and semi-supervised deep learning is a method that combines both weakly supervised and semi-supervised deep learning techniques for the classification of breast ultrasound images [11]. Weakly supervised deep learning is a method in which the model is trained on image-level labels, rather than pixel-level labels. This is useful in situations where obtaining pixel-level labels is difficult or time-consuming [12]. There are several ways in which AI can be used to predict invasive disease-free survival (IDFS) in early-stage breast cancer patients. These include machine learning algorithms, such as neural networks, random forest and decision trees, which can be trained on follow-up clinical data to predict IDFS. These algorithms can learn from the data and identify patterns and risk factors associated with IDFS.

ANALYSIS OF CANCER DETECTION USING ARTIFICIAL INTELLIGENCE

Cancer detection using AI has great potential to improve the accuracy and efficiency of a cancer diagnosis. There are several ways in which AI can be used for cancer detection, including the following:

- Image analysis: AI can be utilized to enhance and analyze medical images like MRI images, CT scans, and x-ray for detection of cancer signs. This can include detecting tumors, identifying malignant tissue, and predicting the stage of cancer.
- Biomarker analysis: AI can be used to analyze biomarkers, such as blood or tissue samples, to detect the presence of cancer. This can include identifying genetic mutations or changes in protein levels that are associated with cancer.
- EHR analysis: AI can be used for identification of patterns and risk factors associated with cancer. This can include identifying patients with high risk of developing cancer and monitoring the progression of cancer in patients who have already been diagnosed.

One of the key advantages of AI-based cancer detection is its efficiency to process large amounts of data. AI algorithms can also learn from the data and improve over time, which can lead to more accurate and consistent diagnoses. However, for using full potential of AI in detection of cancer, there are some challenges to be considered. These include:

- Data quality and availability: AI requires large data to train and test algorithms. Ensuring the availability and quality of this data is a major challenge.

- Explainability: AI-based systems are often difficult to interpret, which can make it difficult for clinicians to understand how and why a diagnosis was made.
- Bias: Biased results can be given by AI models if the trained data are not according to population, which can result in false predictions.

Overall, AI has the potential to significantly improve the accuracy and efficiency of cancer detection, but it is important to address these challenges to ensure that the technology is safe and effective for all patients.

Study of Results in Different Domain for Cancer Detection Using AI

There are a number of studies and research that have been conducted to detect cancer using AI:

- Medical imaging: A study published in 2019 states that an AI model was accurately detecting breast cancer in mammograms with AUC-ROC value of 98% [13]. Another study published in 2020 stated that the AI model predicts lung cancer using CT Scans with an AUC-ROC value of 97%. [14].
- Pathology: A study published in 2018 found that an AI algorithm was able to accurately detect prostate cancer in biopsy slides with an AUC of 0.99 [15]. Another study published in 2019 found that an AI model performed well in detection of ovarian cancer in images with an AUC of 0.98.
- Blood tests: A study published in 2020 stated that an AI model works well in lung cancer detection using blood samples with an AUC of 0.96 [16].

These studies demonstrate that AI can be used to accurately detect a wide range of cancers, including breast, lung, skin, colon, prostate, and ovarian. However, it is important to note that AI-based systems are not yet widely used in clinical practice and are still under evaluation. It is also important to note that these studies were conducted in a controlled setting and the results may not be the same in real-world scenarios. More research is needed to validate these results and to ensure that AI-based systems are safe and effective for use in the clinic.

Analysis on Number of Cancer Cases

According to the World Health Organization (WHO), in 2018 there were nearly 1 crore 81 lakh new registered cases of cancer and 96 lakh deaths caused by cancer. This may increase up to 2 crore 27 lakh new cases and 1 crore 63 lakhs deaths by 2040.

Lung, breast, colorectal, and prostate cancers are the most common types of cancer in the world. In 2018, lung cancer was one of the leading reasons for death in males and females. Breast cancer was the second most deadly cancer, which recorded nearly 6.2 lakh deaths, and colorectal came was recorded as the third most deadly cancer in the world with an estimated 8.8 lakh deaths in 2018. It is important to note

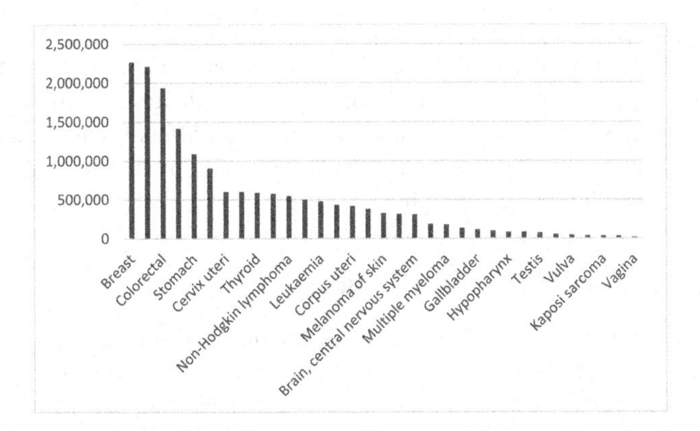

FIGURE 5.3 Number of cases related to cancer [17].

that the number of cancer cases varies widely depending on location and population. The mortality rates of cancer are generally higher in more developed countries, whereas they are lower in less developed countries.

It is also important to note that the number of cancer cases can change over a period of time because of different factors such as advancements in cancer treatments, public health measures, and changes in lifestyle and environmental factors [17].

Lung cancer is the leading cause of cancer death worldwide, and colorectal cancer is the third most deadly cancer globally. Figure 5.3 shows the number of cases related to cancer worldwide and Figure 5.4 shows the rate of survivals in 5 years [18].

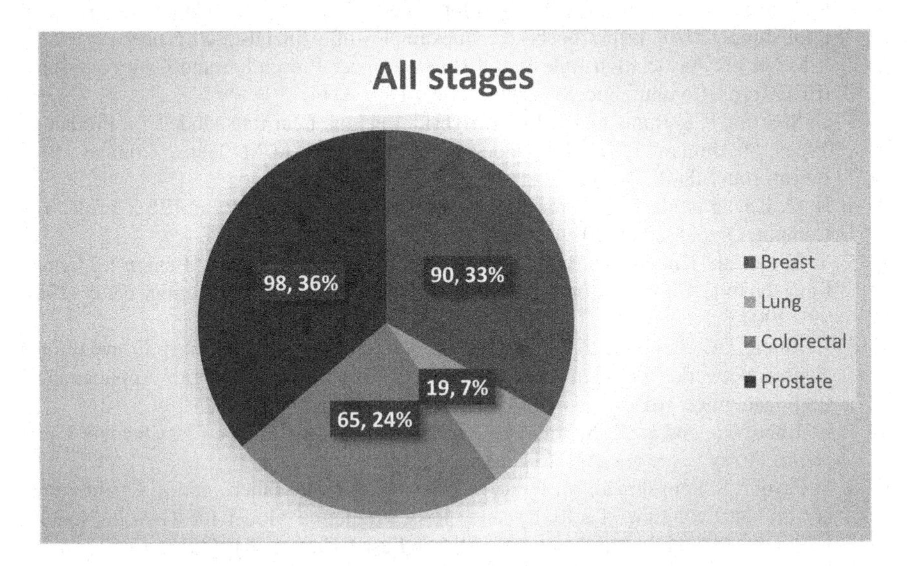

FIGURE 5.4 Five-year survival rate, by stage at diagnosis [37].

The worldwide 5-year standardized survival rate for breast cancer is around 90%. However, there is significant variation in survival rates across different countries and regions, with rates being generally higher in high-income countries and lower in middle- and low-income countries. Factors that can affect breast cancer survival rates include access to early detection and diagnosis, treatment availability, and differences in population demographics.

CONCLUSION AND FUTURE WORK

AI has the capability of providing early detection and suitable treatment of cancer worldwide. Proper analysis of CT scans, MRI images, and pattern and feature identification using AI models can help radiologists and other medical professionals to detect cancer at an earlier stage, when it is more treatable. AI can also be used to identify patterns and risk factors that may indicate an increased likelihood of developing cancer. This can help to improve cancer screening and early detection programs, and to identify populations that are at high risk of developing cancer, so that they can be targeted with preventative measures. Moreover, AI can also be utilized to monitor the progression of cancer and predict response to treatment, which can help to optimize treatment plans and improve outcomes. In conclusion, AI has the potential to revolutionize cancer detection and treatment worldwide, but more research is needed to fully understand and optimize its capabilities. Future work will be based on analyzing technical findings, and the role of AI in curing cancer will be studied.

REFERENCES

1. R. Agarwal, 15 Examples of Artificial Intelligence You're Using in Daily Life, Beebom Publishing, (2020), https://beebom.com/examples-of-artificial-intelligence/.
2. J. Ma et al., Association Rule-Based Breast Cancer Prevention and Control System, IEEE Trans. Comput. Soc. Syst., 6(5) (2019) 1106–1114.
3. G. Sharma, P. S. Rana, and S. Bawa, Hybrid Machine Learning Models for Predicting Types of Human T-Cell Lymphotropic Virus, IEEE/ACM Trans. Comput. Biol. Bioinforma., 18(4) (2019) 1524–1534.
4. H. C. Cheng et al., Deep-Learning-Assisted Volume Visualization, IEEE Trans. Vis. Comput. Graph., 25(2) (2019) 1378–1391.
5. Y. Xie et al., Knowledge-Based Collaborative Deep Learning for Benign-Malignant Lung Nodule Classification on Chest CT, IEEE Trans. Med. Imaging, 38(4) (2019) 991–1004.
6. J. Zhang, A. Saha, Z. Zhu, and M. A. Mazurowski, Hierarchical Convolutional Neural Networks for Segmentation of Breast Tumors in MRI With Application to Radiogenomics, IEEE Trans. Med. Imaging, 38(2) (2019) 435–447.
7. A. Bhardwaj, and K. K. Siddhu, An Approach to Medical Image Classification Using Neuro Fuzzy Logic and ANFIS Classifier, 2013.
8. S. Paydar, S. Pourahmad, M. Azad, S. Bolandparvaz, R. Taheri, and Z. Ghahramani, et al., The Evolution of a Malignancy Risk Prediction Model for Thyroid Nodules Using the Artificial Neural Network. Middle East J. Cancer, 7(1) (2016) 47–52.
9. S. K. Sood, and I. Mahajan. Wearable IoT Sensor-Based Healthcare System for Identifying and Controlling Chikungunya Virus. Comput. Ind., 91 (2017) 33–44.

10. J. Sun, Y. Guo, X. Wang, and Q. Zeng, mHealth for Aging China: Opportunities and Challenges. Aging Dis., 7(1) (2016) 53–67.
11. S. Y. Shin, S. Lee, I. D. Yun, M. Kim, and K. M. Lee., Joint Weakly and Semi-Supervised Deep Learning for Localization and Classification of Masses in Breast Ultrasound Images, IEEE Trans. Med. Imaging, 38(3) (2019) 762–774.
12. B. Fu, P. Liu, J. Lin, L. Deng, K. Hu, and H. Zheng., Predicting Invasive Disease-Free Survival for Early Stage Breast Cancer Patients Using Follow-Up Clinical Data, IEEE Trans. Biomed. Eng., 66(7) (2019) 2053–2064.
13. A. M. Elshaer, and M. A. Elshaer, Artificial Intelligence in Cancer Detection: A Review of the Literature, J. Clin. Oncol., (2021).
14. O. Ozdemir, R. L. Russell, and A. A. Berlin, A 3D Probabilistic Deep Learning System for Detection and Diagnosis of Lung Cancer Using Low-Dose CT Scans, IEEE Trans. Med. Imaging, 39(5) (2020) 1419–1429.
15. J. Chen, and Y. Liu, Deep Learning for Cancer Detection: A Survey, IEEE J. Biomed. Health Inf., (2021).
16. A. Husaini, M. H. Habaebi, A. Hameed, M. R. Islam, and T. S. Gunawan, A Systematic Review of Breast Cancer Detection Using Thermography and Neural Networks, IEEE Access, 8, (2020) 208922–208937, doi: 10.1109/ACCESS.2020.3038817.
17. World Cancer Research Fund International, Worldwide Cancer Statistics, 2020. https://www.wcrf.org/cancer-trends/worldwide-cancer-data/
18. National Cancer Institute, Survival rate, by Stage at Diagnosis, 2020. https://www.cancer.gov/about-cancer/understanding/statistics

6 Ethical Issues in Medical Data

Naga Durga Saile K and B. Venkatesh
VNR Vignana Jyothi Institute of Engineering and Technology, Hyderabad, Telangana, India

INTRODUCTION

Data is a huge collection of information. Information pertaining to the domain of interest is gathered from various sources to form a set of data. With the increase in the areas of machine learning and artificial intelligence, there is always an increased urgency in the datasets of records. With the outburst of pandemic-like issues, cancerous cells, etc., the need for analysis and predictions by the medical field have moved to a greater extent. To generate this medica data as a great deal of personal information is being gathered, there is a set of guidelines, namely ethics, that is to be followed and maintained for the medical data. This chapter discusses the ethical issues that are generated from medical data and ways to maintain ethics.

Medical data is the structured documentation of a patient's medical history and the medication that is given by the healthcare provider to improve the patient's health. Medical records, health charts, and health records are all synonyms for medical data [1]. This medical data comprises the diagnosis, reports generated, and the prescribed medication for the betterment of the patient. There is a history of 4000 years of maintaining medical records [2]. Unlike the latest digital methods, earlier medical records were written on paper by the healthcare provider and these records have been preserved for many years. In the late 1600 BC, the records were kept on a material called papyri. Ancient Egyptians took the stem of the papyri and made sheets so that letters could be inscribed on it [3].

Later, the maintenance of records was done using paper and all the documentation of the records was maintained in hard copies. These medical data are comprised of handwritten prescriptions and case sheets that are not legibly written and not understandable by the common man. The digital era brought an immense change in storing medical records.

In the early years, computers were used for official or financial purposes, but later, with the advent of the Internet and many other programming languages, customized software was incorporated into hospitals to maintain healthcare records. Though digitalization solved the issue of illegible handwriting and understanding the underlying problem with the patient, many other issues needed to be addressed. Studies showed that improper documentation of medical data leads to the misuse of others' data and many instances of data violations have taken place [4]. The key components that are recorded in the Manual or Electronic Health Records are discussed in Table 6.1 [5, 6].

 DOI: 10.1201/9781003388319-6

TABLE 6.1
Key Components of a Medical Record

S. No.	Key Components
1	Name and age
2	Contact details
3	Medical history of family
4	Individual medical history
5	Underlying health details
6	Prescriptions
7	Medications
8	Diagnostics related information (tests and x-rays)
9	Health related policies and Insurances
10	Genetical information

The following sections discuss the rules to be followed legally regarding medical data, and the ethics, concerns, and possible solutions for medical data sharing.

ETHICS FOR MEDICAL DATA

Ethics offer a conceptual method to evaluate and guide morals in decision-making. The word ethics is coined from the Greek word *ethos,* which means a way of living [7]. Ethics in medical data refers to the methodology used and the rules to be followed when using that methodology.

Medical data comprises the information related to a patient or the data of an individual who is ready for clinical trials. Electronic health records are one of the best forms of medical data available. The clinical data comprises the data about the vitals of an individual, diagnostic-related information, and other prescribed information. All these are available in digital form for easy data access and sharing. The ethics to be followed while dealing with medical data are as follows (Figure 6.1) [8]:

1. Respect the consent
2. Privacy of data
3. Confidentiality of identity
4. Data transparency
5. Ownership of the data collected

Respect the Consent

The consent of the person plays an important role in data sharing. If the data of any person are required for research purposes, clinical trials, or any use, before the process is initiated, we must take the consent of the individual and help them understand the underlying purpose for the being requested, and the importance of data sharing. If the individual is reluctant to share the data, assurance of data confidentiality and privacy should be given. Without proper consent, it is illegal to use the data for any

1. Respect the consent
2. Privacy of data
3. Confidentiality of identity
4. Data transparency
5. Ownership of the data collected

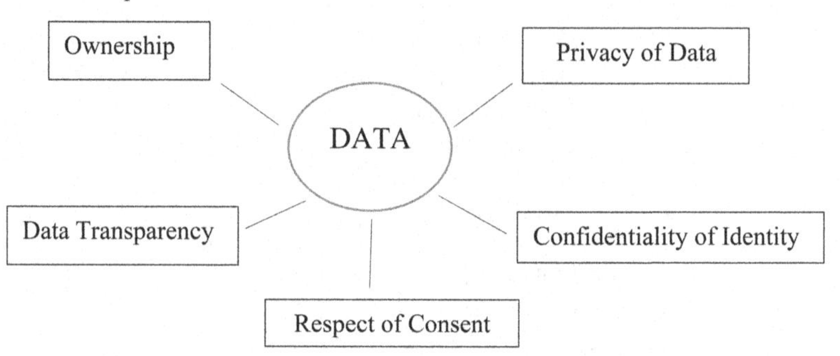

FIGURE 6.1 Ethics of medical data.

kind of activity. There are a few cases where the consent can be waived off by the Institute Review Board (IRB) [9] and data can be made available. The IRB thoroughly verifies the purpose and prejudice in sharing the data.

Privacy of Data

A huge proportion of people argue for a right to privacy of an individual's data, particularly regarding health details. Be it physical health or mental health it is the utmost responsibility of doctors and hospitals to maintain the privacy of one's data [10]. As per the Right to Information Act 2005, in India, it is the right of every individual that his or her data is maintained confidentially [11]. The Act says that until and unless the data are required for legal proceedings or by any government organization, privacy and confidentiality are to be achieved. If any organization or individual breaks the law, they can be sued. As per the regulations of the Indian Medical Council of India, any medical professional is obligated to maintain the confidentiality of the patient's information and failing to do so the personnel is considered guilty of the event [12]. It is also observed as misconduct in the behavior of the professional who misuses the patient record [12]. The privacy of data can be achieved by following protocols for maintaining and data sharing as follows:

- Storing data in a secured format
- Providing restricted access permission for the staff
- Restricting the visibility of the data to all
- Providing encrypted password protection of the document
- Sharing only the required data

The electronic health records should be stored in a secured data format. The file access permissions like editing, modifying, and viewing should be constrained and restricted among the staff. Whoever accesses the records should have a unique ID and password and the purpose the records are being accessed needs to be documented. The person who is in charge of the recordkeeping should provide the health records with password encryption to avoid unauthorized user access. The most important thing is to share only the data that are required, rather than dumping it with unnecessary data. Most of the time people tend to share the entire file of an individual, which leads to unnecessary data transfer, rather than keeping track of the requirements and scrutiny of what kind of data is required and for what purpose (this needs to be identified the earlier stage of sharing the data).

Confidentiality of Identity

The identity of the individual should be kept confidential. Identity includes name, address, personal identification number, face, etc. The confidentiality of identity can be achieved by encrypting the name with dummy variables, addresses, and identification numbers by using alphanumeric [13] and other encryption methods. The face of an individual can be hidden by hiding it completely with a blank space, as displayed in Figure 6.2, or blur the image completely.

The most common technique for maintaining confidentiality of identity is hiding the eyes with a blank space, as shown in Figure 6.2. These techniques help the individual to gain confidence that their identity remains confidential. This is one of the important ethics to be followed in medical data.

Data Transparency

Medical records are to be provided with proper transparency of data. In an electronic medical record (EMR) not all the fields are required for a researcher. Only the data that are mandatory for research needs to be shared to maintain privacy of data. Once the research or clinical trials are completed, the results or the conclusions should be shared with the authorities for final approval before completion. If this transparency of the work is not available, it may lead to devasting effects on human life. In one instance, a drug was not scrutinized by the World Health Organization (WHO) before approving it for real-world usage. Because of the lack of proper data transparency the drug was used and caused ill effects [14] on human health. In another case,

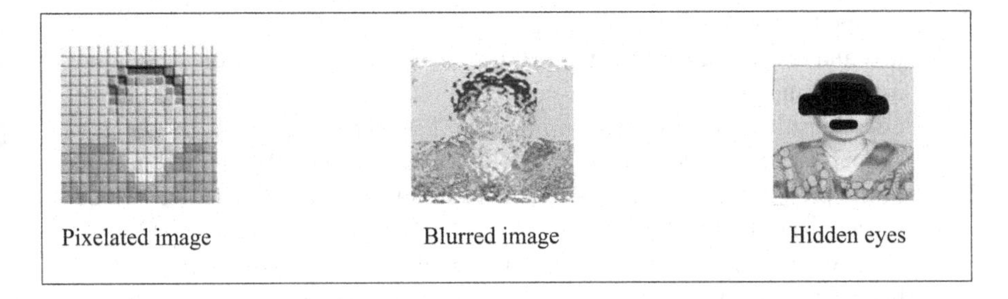

FIGURE 6.2 Different techniques use to ensure confidentiality of identity.

an anti-diabetic drug that was manufactured to treat cardiovascular effects increased the rate of deaths and heart attacks because the drug was not scrutinized properly before release because of inadequate data transparency. Any statistics of the data or research is to be made transparently available so that governments and other bodies can take necessary actions during situations like the pandemic. As we studied that privacy and confidentiality of data are important, similarly after certain research the transparency of the outcomes and the statistics of the data are equally important [15]. Improving the quality of transparency and reporting is one of the major ethics in medical data research.

Ownership of the Data Collected

Ownership is an important ethic in terms of medical data. The whole and sole responsibility of the collected data lies with the person who has gathered it. The individual should realize that he or she is the owner of the data, and that the data has not been misused (theft or loss of information). We all well know that privacy and confidentiality can be assured only with proper ownership of the data collected. Sharing of the data with multiple people is illegal or unethical. Apart from serving the purpose of the data, using or manipulating the data for any other purposes is considered completely to be unethical.

POLICIES AND LAWS

World Health Organization

Data is the major ingredient for public health, and the benefits of data sharing help in leveraging better outcomes through research. Research has been globalized and there are a great deal of research activities happening across the globe in various areas of medicine. With the advent of big data and analytics, research has taken a plunge. When gathering data for various purposes, privacy and confidentiality of the data have always been a major concern. To achieve proper data sharing, the WHO [16] has created a policy regarding the data it collects and shares. The policy states that the member states and WHO shall do the following:

- Use the rule of data privacy and confidentiality of the data collected.
- Use and publish any data of an individual only if the identity is to be stripped off.
- Use and disclose the results to WHO so there is a unique way of accumulating data analytics.
- Use measures for ethical and secure use of the data.
- As per the ISO 27001 standard [16], the employees of WHO shall adhere to the policy of providing data security in all forms of applications, cloud storage, and the Internet.
- To safeguard all the activities of data, ensure a review committee is established to monitor the activities of the data, and keep keen monitoring of data circulation.

HIPAA in the United States

As per a survey, it is reported that 15% of doctors say that maintaining the privacy and confidentiality of a patient is the highest ethical and prominent underlying practice [17]. Any malpractice of this leads to termination or suspension as per the Health Insurance Portability and Accountability Act of 1996 (HIPAA).

HIPAA is a privacy law created by the federal government of the United States that was initiated to protect sensitive patient-related data being shared without the consent of the individual. These privacy rules comprise a few entities that must adhere to the policies of the act. The entities are known as covered entities [18] and are as follows:

- Health care providers
- Health insurance providers
- Business associates
- Healthcare clearing houses

The data privacy and confidentiality of the patient's information as per HIPAA are confronted only by these entities. *Healthcare providers* are the hospitals and the officials who interact with the patients and help in generating claims. *Health insurance providers* are the organizations and people who demonstrate various methodologies to convince people to attain health insurance plans and understand what the plan entails. *Business associates* are the people who perform the data analysis, claims to preprocess, etc. *Healthcare clearing houses* are the mediators who maintain changing the nonstandard information to a standard information format or vice versa for proper data channeling. As per HIPAA, the health data and EMR are approachable and accessible for these entities with a particular group of permissions.

Permitted Uses and Disclosures

Consent for the data is mandatory to disclose information to third-party users. The data can be provided without the consent of the patient only during the following situations:

- When the data are required by the law
- Law enforcement
- Research under some determined situations
- Organ, eye, or tissue donations
- Government authorities
- Domestic violence
- Reduce a major health risk, etc.

During these scenarios, the consent of the individual is not necessary and the data can be disclosed and used by the authorities. To act under HIPAA, the medical data or EMR or electronic medical history (EMH) should be maintained with

- Confidentiality
- Integrity

- Availability
- Avoid impermissible acts and disclosures
- Detects and safeguards threats of data theft

GDPR in Europe

General Data Protection and Regulation (GDPR) is used to protect data and harmonize privacy in all data-related issues across Europe. The articles under GDPR cover data about health, personal data, ethnic origin, data related to politics, membership details, etc. Processing of data for health is acceptable if the requirements of GDPR Article 9(2) are met [19].

- As per the article, the processing of medical data can be done for preventive medicine generation, accessing the working of the employee, performing diagnosis of a particular disease, etc.
- Special interest in a particular medical field for which data are required.
- The data can be accessed for historical or scientific research purposes or statistical purposes as per Article 89(1) [20].
- As long as the privacy and confidentiality of the data are maintained the data can be accessed.
- Though there is a data protection and regulation act, there are still some states that don't allow medical data, personal data, and biometric data to be shared as per the law of the state.

ICMR Guidelines

Indian Council of Medical Research (ICMR) is a body that governs all the medical and biomedical-related research activities in the country. It is one of the oldest medical bodies performing medical research in various topologies.

As research needs data and medical data requires a number of parameters to perform the data analysis, ICMR has released guidelines to deal with the ethical issues in sharing or using medical data [21]:

- With the increase in data and big data analytics, the need for data and datasets also increased a great deal. Before the medical data are shared, the ethics committee has to review the "whats" and "whys" of the data.
- There are many data mining tools available to get electronic data. If the medical data need to be mined or retrieved the approval of the ethics committee is required. Without prior approval, the data cannot be retrieved.
- Privacy, security, accuracy, and legal liability of the data should be maintained for the outsourced data.
- Due to the rise in data availability, there are a great deal of misuses and misleading of the data. Much of the data are being shared with third-party entities for commercial and business purposes. To overcome this, there are a number of ways to maintain the privacy of data and sharing rights and a real need for the data to be truly scrutinized and reviewed.

- Because there is an increased amount of electronic data and Manual of electronic data and manual health record, utmost care is to be taken for the privacy and confidentiality of the data.

Unlike other countries, India doesn't have a specific data protection and regulations act as the ICMR has given the following guidelines that can be followed for medical data sharing:

1. To make a certain commitment to the physical security of the data.
2. Make the data password protected.
3. Keep a check on the access of the data by the people.
4. Authorized login access to the data is to be made mandatory.
5. When the data needs to be transferred from one location to another make sure there is certain encryption done during the file transfer.
6. Because there are no specific acts, it is always good to ensure the possible legal aspects if the data are being compromised or misled to the person of consent.

The goal for any kind of technology enhancements, whichever it may be, is maintaining the confidentiality and privacy of the data gathered and making proper use of the collected data for the enrichment of the research work.

LITERATURE SURVEY

Existing Frameworks for Maintaining Medical Data Ethics of Privacy and Confidentiality

To achieve privacy and confidentiality of medical data researchers have been working in different areas using the latest technologies of machine learning, Internet of Things (IoT), blockchain, etc. The authors in [22] proposed an architecture for medical data sharing from a clearing house to physicians using blockchain technology. They have proposed a secured hash method of transferring data using a cryptographic hash function from one block to another block. This method ensures data integrity over sharing on the Internet.

The authors in [23] demonstrated an architecture for sharing medical data to attain privacy on the cloud using three approaches to data sharing. The first model is the centralized approach from the main data analysis cell to individual-level data. A second model is a decentralized approach (site-level meta-analysis), and the third one is a decentralized federated learning approach in which iterative global-level data are being used. The authors of [24] proposed a system architecture for encrypting the data in the cloud using an elliptical cryptographic technique using edge server and blockchain technology where the medical data is encrypted using a smart medical card for credentials and edge server along with blockchain hash codes and encryption techniques (the data will be encrypted).

The paper in [25] is research work done to attain the privacy of data by using hybrid cryptographic techniques. The author uses an adaptive elliptical curve algorithm (AECC) along with advanced encryption standard (AES) to provide encryption to the personal health data of an individual. The hybrid model demonstrated shows a good level of encryption for data security and privacy. There is another architecture that was designed to facilitate a great level of privacy of health data stored in the cloud [26], which uses the ElGamal encryption algorithm to attain privacy over the cloud.

The authors in [27] discuss the encryption techniques that use the RSA algorithm for sharing the sensitive health-related information. Using the RSA techniques, the data of every individual patient are encrypted and secure data transmission techniques are used for data in the cloud. These are some of the techniques that are used to maintain better privacy and confidentiality of medical data. Many studies are still being researched regarding the maintenance of security of personal health data.

CASE STUDY

Ethics Committee Guidelines for COVID-19

In the previous sections we discussed the different guidelines and mechanisms that are followed in different situations. As we are well aware of the latest pandemic COVID-19, considering it as a case study, we shall see what guidelines are set up by the ethics committee for COVID-19. Based on the different pandemics, there are always certain ethics that are followed by different governments. Now, let us see the guidelines that are maintained for the medical data being used for research on COVID-19 [28] issued by ICMR.

- The very basic and important principles of ethics are autonomy, beneficence, non-maleficence, and justice. These four principles play a very important role in maintaining the privacy and integrity of the data.
- It is the responsibility of the researcher to gain the trust of the ethics committee so that there would not be any violations in attaining the security of the data.
- The storage of data and the datasets of the pandemic gave a path to much research work to be carried out on various aspects of human behavior and ideology [29].
- The samples collected and stored should be made anonymous and the post-research benefits need to be shared with the ethics committee.
- Appropriate custodianship of the data is to be availed from the Ethics Committee (EC) and written consent is to be given for assurance of maintaining the privacy and confidentiality of the data.
- Any kind of research activity that involves collaborations with other countries needs to sign a Memorandum of Understanding on consent issued to maintain the security of the data.
- The ethics committee timely reviews the progress of research activities and it keeps monitoring the data.
- Any protocol deviations would lead to legal action. Hence, the utmost care is to be taken for these types of medical data handling.

CONCLUSION

Medical data are highly confidential, and require privacy and ownership. Ethical responsibility of the medical data lies in the hands of the concern who are using it. Though different nations have different guidelines for medical data sharing, all adhere to the umbrella rules designed by the WHO. Ethics that are properly maintained and well formulated gives rise to a smaller number of legal issues. Because there is a lot of personal information being shared in medical data, it is also the responsibility of the researchers to adhere to the guidelines of the ethical review committee and maintain the utmost privacy, confidentiality, and integrity of the medical data. Thus, we can say that by using appropriate privacy mechanisms like encryptions we can maintain the privacy of data and keep the ethics in medical data usage intact.

ACKNOWLEDGMENT

I would like to thank Dr. K. Phani Koteswara Rao, APVVP Commissioner (Retd), Medical Superintendent at Basavatarakam Indo-American Cancer Hospital & Research Institute, Hyderabad, for sharing his insights and helping me in drafting the chapter by sharing knowledge and his experiences.

REFERENCES

1. https://en.wikipedia.org/wiki/Medical_record
2. Gillum RF. From papyrus to the electronic tablet: a brief history of the clinical medical record with lessons for the digital age. Am J Med 2013 Oct;126(10):853–857. doi: 10.1016/j.amjmed.2013.03.024. https://pubmed.ncbi.nlm.nih.gov/24054954/
3. Evans RS. Electronic health records: then, now, and in the future. Year Med Inform 2018 Mar 06;25(S 01):48–61. doi:10.15265/iys-2016-s006. https://www.ncbi.nlm.nih.gov/pmc/articles/PMC5171496/
4. Institute of Medicine, Committee on Improving the Patient Record, Detmer DE, Steen EB, Dick RS, editors. The Computer-Based Patient Record: An Essential Technology for Health Care, Revised Edition. Washington, D.C.: National Academies Press; 1997:1–256. https://pubmed.ncbi.nlm.nih.gov/25121222/
5. American Retrieval. The Components of a Medical Record. https://americanretrieval.com/medical-record-components/
6. SoftClinic Genx. Our Blog. https://www.softclinicsoftware.com/what-are-key-components-of-a-medical-record-or-electronic-medical-records/
7. Government of Canada. What is Ethics? https://www.canada.ca/en/treasury-board-secretariat/services/values-ethics/code/what-is-ethics.html
8. Global Health Data Methods. Ethics of Collecting and Using Health Data. https://globalhealthdata.org/ethics-of-collecting-and-using-health-data/
9. Gliklich RE, Dreyer NA, Leavy MB, editors. 7. Principles of Registry Ethics, Data Ownership, and Privacy. Registries for Evaluating Patient Outcomes: A User's Guide [Internet]. 3rd edition. Rockville (MD): Agency for Healthcare Research and Quality (US); 2014 Apr. https://www.ncbi.nlm.nih.gov/books/NBK208620/
10. Hartlev M. Striking the right balance: patient's rights and opposing interests with regard to health information. Eur J Health Law. 2007 Jul;14(2):165–176. doi: 10.1163/092902707x199122. https://pubmed.ncbi.nlm.nih.gov/17847831/

11. Ministry of Law and Justice, Government of India. The Right to Information Act, 2005. The Gazette of India, Part II, Section 1. New Delhi: Government of India; 2005, Jun 21. https://rti.gov.in/rti-act.pdf
12. Indian Medical Council. Professional Conduct, Etiquette and Ethics Regulations, 2002. https://wbconsumers.gov.in/writereaddata/ACT%20&%20RULES/Relevant%20 Act%20&%20Rules/Code%20of%20Medical%20Ethics%20Regulations.pdf
13. Biotechnology Innovation Organization. Confidentiality of Patient Medical Records. https://archive.bio.org/articles/confidentiality-patient-medical-records
14. Joshi M, Bhardwaj P. Impact of data transparency: Scientific publications. Perspect Clin Res. 2018;9(1):31–36. https://www.ncbi.nlm.nih.gov/pmc/articles/PMC5799949/
15. Nicholls SG, Langan SM, Benchimol, EI, Moher D. Reporting transparency: making the ethical mandate explicit. BMC Med. 2016;14:44. https://bmcmedicine.biomedcentral.com/articles/10.1186/s12916-016-0587-5
16. World Health Organization. WHO Policy on the Use and Sharing of Data Collected by WHO in Member States Outside the Context of Public Health Emergencies. https://www.who.int/about/policies/publishing/data-policy
17. Physician Resources. Identifying and Navigating Ethical Issues in Healthcare. https://www.sermo.com/resources/ethical-issues-in-healthcare/
18. Centers for Disease Control and Prevention. Health Insurance Portability and Accountability Act of 1996 (HIPAA). https://www.cdc.gov/phlp/publications/topic/hipaa.html
19. Intersoft Consulting. Art. 9 GDPR: Processing of Special Categories of Personal Data. https://gdpr-info.eu/art-9-gdpr/
20. Intersoft Consulting. Art. 89 GDPR: Safeguards and Derogations Relating to Processing for Archiving Purposes in the Public Interest, Scientific or Historical Research Purposes or Statistical Purposes. https://gdpr-info.eu/art-89-gdpr/
21. Indian Council of Medical Research. National Ethical Guidelines for Biomedical and Health Research Involving Human Participatnts. https://main.icmr.nic.in/sites/default/files/guidelines/ICMR_Ethical_Guidelines_2017.pdf (11.7)
22. Patel V. A framework for secure and decentralized sharing of medical imaging data via blockchain consensus. Health Inf J. 2019;25(4):1398–1411. doi: 10.1177/1460458218769699. https://journals.sagepub.com/doi/pdf/10.1177/1460458218769699
23. Scheibner J, Raisaro JL, Troncoso-Pastoriza JR, Ienca M, Fellay J, Vayena E, Hubaux JP. Revolutionizing medical data sharing using advanced privacy-enhancing technologies: technical, legal, and ethical synthesis. J Med Internet Res. 2021 Feb 25;23(2):e25120. doi: 10.2196/25120. https://pubmed.ncbi.nlm.nih.gov/33629963/
24. Christo MS, Jesi VE, Priyadarsini U, Anbarasu V, Venugopal H, Karuppiah M. Ensuring improved security in medical data using ECC and blockchain technology with edge devices. Secur Commun Netw. 2021;2021:6966206. https://doi.org/10.1155/2021/6966206
25. Madhavi M, Sasirooba T, Kumar GK et al. SRMDS: Secure and Reliable Medical Data Sharing Using Hybrid Cryptography with Block Chain in E-Health Systems, 05 October 2022. https://doi.org/10.21203/rs.3.rs-2067301/v1
26. Thilakanathan D, Calvo RA, Chen S, Nepal S, Glozier N. Facilitating secure sharing of personal health data in the cloud. JMIR Med Inform. 2016 May 27;4(2):e15. https://medinform.jmir.org/2016/2/e15/
27. Sharma K, Agrawal A, Pandey D,. Khan RA, Dinkar SK. RSA based encryption approach for preserving confidentiality of big data. J King Saud Univ. 2022;34(5):2088. https://doi.org/10.1016/j.jksuci.2019.10.006.
28. EC_Guidance_COVID19.pdf (ncdirindia.org) https://ethics.ncdirindia.org/asset/pdf/EC_Guidance_COVID19.pdf
29. Saile KND, Navatha K, Kavitha A. Sentiment analysis on lockdown Covid-19. Int J Indian Psychol. 2022;8(2). https://ijip.in/wp-content/uploads/2020/06/18.01.057.20200802.pdf

7 AI-Based Waste Classification in the Healthcare Industry

Saswati Kumari Behera, Aouthithiye Barathwaj SR Y, Vasundhara L, Saisudha G, and Haariharan N C
Sri Sai Ram Engineering College,
Chennai, Tamil Nadu, India

INTRODUCTION

The collection, storage, categorization, and transportation of solid wastes are all part of the waste management process, especially in the healthcare industry. While recovering from a serious illness or surgery, diet is crucial. Based on the needs of the patient, all reputable hospitals offer a balanced diet that is nutritious and healthy. For patients who are hospitalized for short or long stays, the hospital catering services are essential in supplying appropriate diet food. Hospital catering provides a great deal of vegetable wastes that are generated every day. People drink water in the bottle and throw the bottles in the dustbin. Also, everyday a huge amount of biological waste is generated in the healthcare industry. All these types of waste are all included in waste management. Planning, instruction, and training are essential for each of these processes. The waste is primarily produced by human activity. Glass, paper, plastics, wood, and vegetable waste are among the most typical types of waste. A significant role is played by waste classification, which is a time-consuming procedure, particularly in larger cities. Waste management varies greatly all over the world, with each country having its own approach; it is vital for living in a sustainable manner. A clean, hygienic, and pleasant environment necessitates proper waste management. People are currently engaged to manually segregate waste (hand-picking method) [1]. This necessitates a tremendous amount of human effort. Workers are at risk because they physically segregate garbage, exposing themselves to the risk of infection from germs found in the solid waste. It is prohibitively expensive to separate waste the traditional way. Municipal solid waste (MSW) is typically disposed of in landfills and bodies of water. Groundwater and surface waters are both contaminated by garbage dumped in landfills. In crowded cities especially, underground water is a crucial source of water for people, putting them in danger when it becomes contaminated. Batteries and lead materials at the landfill primarily cause soil pollution. According to estimates, there are roughly 10,000 metric tons of MSW produced annually. The environment is at risk if these wastes are not properly disposed of and separated. As soon as possible, waste should be separated into different

DOI: 10.1201/9781003388319-7

categories to increase the proportion of recycled and reused materials and decrease litter contamination. By separating waste, disposal costs are decreased. Additionally, waste segmentation enables us to use the best waste treatment and disposal practices. Incorrect garbage disposal, particularly with regard to common household waste, is to blame for about 75% of infectious diseases. Mixing of waste with used water from households and other elements led to the spread of typhoid, cholera, and tuberculosis. On the other hand, the air may become contaminated when these mixed/non-segregated wastes are burned. In the worst-case scenario, cancer could result from incineration of these wastes if they contain any hazardous materials [2]. This is mainly because when these wastes are burned, tiny harmful particles are released. Deep learning and computer vision are fields that are rapidly developing, especially in the area of waste management. Furthermore, as image processing techniques and architectures advance quickly, we are forced to focus our efforts on meeting the growing demand for waste classification. To recognize and categorize the recorded image data, convolutional neural networks (CNNs) are used. Waste is divided into recognizable groups and types using support vector machines (SVMs). In contrast to the traditional segregation method, this method is significantly more accurate and efficient than hand-picking [3].

RESEARCH METHODOLOGY

Advantages and Advancements of AI in Image Classification

Artificial Intelligence (AI) is a digital mind that processes, evaluates, and examines the data that is fed into the system. AI aims to simulate a human lifestyle with better reasoning and perception. It proves more efficient to make the right decisions faster.

With the recent advancements in this field, along with image manipulation, AI would overtake manual efforts with better understanding of data and accuracy. Machine learning is used to process and analyze images and AI recognizes the size, shape, position, and color. It consists of various algorithms that are used for manipulation of the images. The major aspects of image manipulation include image analysis and image segmentation, neural networks, SVMs, and a Gaussian mixture model [4]. Initially, the image is captured with the help of a sensor. Further, the quality of image is enhanced and corruption, if any, is also removed to acquire the mathematical model basis. RGB processing can be used for color images. Similarly, morphological processing processes the image shape and structure. AI and ML together increase the speed and efficiency of the process when large data is fed into the system. This combined effort of AI and ML also decreases the manual effort that is required for processing every single data.

Computer vision deals with training of system to recognize and process the images in the real world. Neural networks and deep learning are the tools that are used for processing complex image models. There are various popular neural networks for complex image processing. The most common network type is the CNN. This is adapted for image processing combined with AI, and it is a class of deep learning networks. The structure of CNNs is much similar to the neuron structure in the human brain. This multilayer network has various layers for image manipulation. The convolutional layer filters and abstracts the fed-in image. Pooling layers remove the corruptions in the image. Normalization layers are used to improve the network's performance. The last layer, fully connected layer, joins the previous levels to obtain input and produce output.

Various implementations of AI algorithm with deep learning and neural networks opens up a wide door for large technical advancements in the world.

Applications:

1. Fingerprint voice identification technology

    Fingerprint identification technology deals with the uniqueness of the fingerprint. This technology mainly concentrates on the color fingerprint that allows us to identify a person's identity. The voice identification technology groups the voices on different age groups based on sound and information of the data.
2. Face recognition technology

    Face recognition technology is a biometric identification system that relates the biometric facial pattern to identify a person's identity. This technology proves to be more accurate because its deals with pupil features that are highly unique.
3. Radiofrequency ID (RFID) and barcode recognition technology

    RFID involves electromagnetic waves to track and transfer data through a transmitter and receiver antenna. A barcode identification system includes QR tag techniques.

Waste Classification and Its Importance

Domestic waste management is one of the most pressing concerns facing environmental governance today. Residents are the primary producers of household garbage, and the problem is exacerbated in places with high populations and rapid consumption.

People's increasingly fulfilling material lives generate a rising amount of domestic garbage, which comes in a variety of forms. Although this demonstrates that our living conditions are rising, the impact of waste on our lives is also increasing. The worse the problem gets, the more vital it becomes to figure out how to better deal with domestic garbage. One of the most effective approaches is a domestic garbage classification or sorting system.

Using a waste sorting system or waste categorization is the best strategy and best way to combat waste pollution. Garbage classification is now an essential step in a nation's development. By classifying garbage, people can learn how to preserve resources, use resources wisely, develop good living habits, and enhance their own quality of life. Effective waste sorting habits will help someone become more aware of environmental issues, understand the value of life's resources, and develop the habit of resource conservation.

Chemicals are present in some modern waste, some of which, if unidentified, will lead to an increase in human incidence. Even if you landfill the garbage far from your home and use the necessary isolation technology, it is difficult to prevent the penetration of harmful substances that will enter the entire ecosystem with the Earth's circulation. The health of people will eventually be impacted by plant and animal pollution of soil and water supplies.

Waste classification is the process of grouping different types of trash into distinct categories according to the impact that each type's use has on the environment, the makeup of the trash, and the needs of the treatment technologies that are currently in use. Garbage classification serves the following purposes:

1. Separate the kitchen wastes, which are primarily composed of perishable organic ingredients, to produce high-quality raw materials for garbage composting and high-quality organic fertilizers, which will help to improve soil fertility and decrease fertilizer use.
2. By designating waste as hazardous, it is possible to treat waste without harming the environment and lower the danger of air, water, and soil pollution. Hazardous waste is waste that has trace amounts of pathogenic microorganisms, organic pollutants, and heavy metals.
3. Increase the percentage of waste that is recycled, lower the demand for raw materials, and lower CO_2 emissions.

Classifiers in Practice

Waste products are currently classified using a variety of techniques. Based on the material's density, waste is categorized using x-rays. The waste is sorted according to object size in a trommel separator, which consists of a rotating drum with perforations. With this technique, the drum is used to separate smaller waste particles from larger waste particles and collect each type separately. Large-scale induction sorting is a common practice for waste classification. This method involves placing and moving the waste along a conveyor belt. The conveyor belt is equipped with a number of sensors that detect waste and categorize it as metal, wood, plastic, and other materials.

Pressured air jets installed along the conveyor belt are then used to move these wastes to the proper collector bin. We make use of the electromagnetic theory to classify waste. Eddy current separator is the name given to this type of segregator, which is typically used to separate ferrous from non-ferrous waste when dealing with metallic waste. Because this Eddy current separator is primarily used for waste classification in metallic-based industries, it is not recommended for non-profit use. Near-infrared (NIR) sensors are the other type of segregator and, as of now, are the most successful in both profit and non-profit applications [5].

This technique categorizes waste objects according to their reflectance characteristics. Of all the waste classification techniques, this one is the most successful [6]. The ability to identify and categorize differs greatly from other methods because various materials have different reflectance characteristics. The results, however, may be incredibly unpredictable if one type of waste is piled on top of another.

Data Collection

In the field of object categorization using images, the dataset is absolutely crucial. To achieve the desired results, it is crucial to draw the appropriate conclusions from the supported fact data mining. It is defined as the effective application of large datasets using mathematical and statistical techniques. Image processing frequently employs MATLAB and OpenCV. In the dataset, there are 470 images. The information was specifically gathered to take into account the many factors that affect waste sorting. Other factors include mixed waste, overlaying, uneven illumination, and others.

Preprocessing of images is carried out to accommodate these discrepancies. To carry out data mining, we employ SVMs [7, 8] in the suggested strategy. With SVMs, classification and regression problems can be resolved. Each feature and its associated value are displayed by an SVM at a different coordinate in an n-dimensional space. Hyperplanes are divided according to the multidimensional information they hold, and this serves as the basis for classification.

$$\begin{gathered} \min_{y,w,b} \frac{1}{2}\|w\|^2 \\ y^{(i)}\left(w^T x^{(i)} + b\right) \geq 1, i = 1 \ldots m \end{gathered} \tag{7.1}$$

We use linear SVM for the specified classification. The optimization of SVM is represented by Eq. (7.1), where y^i is is the cost function, w^T is the weights, b is bias, and i is the sample range from 1 to m, because there are more features in the image dataset than observations. SVMs are particularly effective for higher dimensional data. SVM decreases memory usage during image pre-processing. The strength of the decision function is also enhanced by the use of various kernel functions.

YOLOv4

CNN supports the YOLO architecture. The CNN separates an image into regions before predicting the boundary boxes and probabilities for each region. It forecasts a number of bounding boxes and probabilities for each class at the same time. YOLO implicitly encodes context in addition to visual information about classes because it sees the entire image during training and testing.

The majority of current accurate models are extremely slow and inefficient when trained with a small mini-batch size on a single GPU, and require multiple GPUs. By developing an object detector that can be trained on a single GPU with a smaller mini-batch size, this problem is addressed in YOLOv4. This enables the training of an extremely quick and precise object detector on a single 1080 Ti or 2080 Ti GPU.

Some backbones are better suited to classification than detection. For instance, CSPResNext50 outperformed CSPDarknet53 in terms of im-age categorization, but CSPDarknet53 outperformed CSPResNext50 in terms of object detection. According to the article, a backbone model for object detection needs more layers for a larger receptive field and a larger input network size for better identification of small objects. An extractor of features, A backbone is a deep learning framework.

Every backbone model is essentially a classification model. I'm going to assume that everyone at least knows about VGG16, one of the earliest deep learning classification systems. SqueezeNet, MobileNet, and ShuffleNet are three additional backbone models in addition to the ones mentioned earlier.

A subset of the bag of specialties called neck is made up of feature maps from different backbone phases; simply put, it is a feature aggregator. The neck of the object detection pipeline will be more thoroughly discussed in the following subsection.

The object detector, also referred to as the head, locates a potential object location but cannot identify the specific object that may be present. There are anchor-based and anchor-free detectors as well as two-stage and one-stage detectors available. The head will be covered in more detail in the sections that follow. YOLOv4 applies SAT to the training set, perturbing the data by an amount X until the predicted label matches the original class. This assists in the model's generalization.

The recognition of handwritten, printed, or typed characters from an image is known as optical character recognition (OCR). These methods are used to convert printed books to digital files. Data entry, traffic sign recognition, and other applications are examples.

Self-driving cars are vehicles that can drive themselves. Detecting pedestrians, automobiles, trucks, traffic signs, and other objects some of self-driving cars' most important capabilities. These detections are necessary for self-driving automobiles to function properly.

In iPhone and Android phones, face and IRIS verification and authentication are employed. If an exact face or iris match is detected, the device is authorized.

Object detection has a wide range of applications in robotics. Bin selecting and sorting of objects is one of the most prevalent applications. The robot can understand the location of objects using object detection algorithms. The robot will be able to pick up and sort the object using this information.

You can monitor an object and utilize it as an object counter by using object detection algorithms. For instance, how many automobiles have crossed at a crosswalk, how many people have entered a retail mall, and so forth

EXPERIMENT AND ANALYSIS

The proposed system consists of two compartments for biodegradable and non-biodegradable wastes, a servo motor, a disc for auxiliary lid rotation, a Raspberry Pi, a Raspberry Pi high-quality camera, and all of these parts put together. Additionally, a chamber for quick composting of biodegradable wastes has been built beneath this structure.

The suggested system includes a Pi, a high-resolution Pi camera, a motor, a disc for secondary closure rotation, and dual sections for recyclable and non-recyclable waste are included. Furthermore, a chamber beneath this structure is constructed for the fast composting of biodegradable wastes. Once dumped into the bin, the waste drops to the secondary lid below. The Raspberry Pi's high-resolution camera then captures the waste in low illuminance due to its interchangeable lenses that can acquire high-quality image data even in dark conditions. This image is then analyzed and classified by YOLOv4 weights that have been trained using explicit

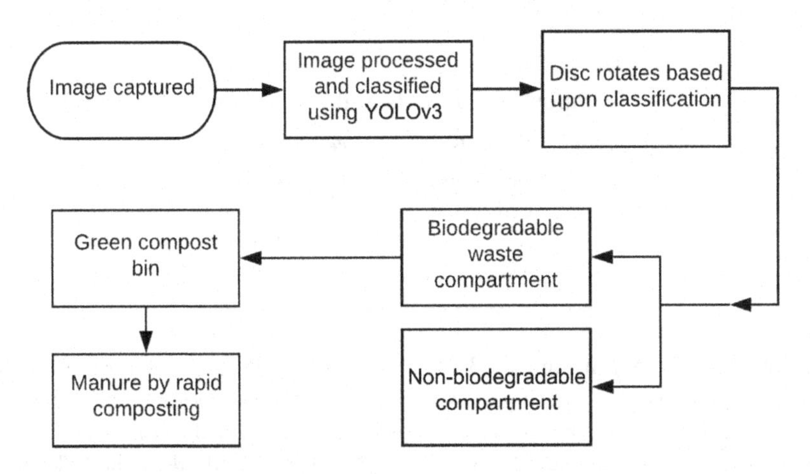

FIGURE 7.1 Flow of control and data in the proposed system.

data [9–14], and based on their classes, the disc rotates the auxiliary cover to direct it to its respective compartment. Afterward, biodegradable waste is shifted to a chamber devoted to quick composting according to Berkeley's quick composting method [15–19], which involves thermal composting at a temperature of 55–65°C to maintain C:N balance at 30:1. Activators such as animal urine, nettles, old compost, or other natural activators are used for the decomposition process, which can take up to 14 days before converting waste into manure [20]. Furthermore machine learning algorithms such as SVMs are used for preprocessed image classification with the YOLOv4 technique with 470 photos divided into two categories (biodegradable and non-biodegradable). The dataset is labeled with LABELIMG, whereas Google Colaboratory is used for training with the maximum batch size set at 2500 and mean average precision (MAP) determined every 100 iterations.

Garbage that is thrown into the bin travels through the primary opening and under the secondary lid. The high-resolution camera on the Raspberry Pi then records the trash. Even in low-light we can record high-resolution image situations thanks to this camera's interchangeable lenses. The control flow is given in below Figure 7.1.

RESULTS

The results are significantly more accurate and efficient compared with training the YOLOv4 algorithm on the data without any picture preprocessing. In addition to increasing detection accuracy, image preprocessing with SMVs reduces memory usage. The percentages of organic material, carbon (C), nitrogen (N), and ash content decomposition after 14 days are displayed in Table 7.1. This novel technique generates more nitrogen and carbon content when compared with cold composting. The process of removing the weeds from the garden was a success.

The precision was able to increase the effectiveness of waste classification in the suggested system by utilizing deep learning, and computer vision could be observed in Figure 7.2. In addition to lowering the cost, this lessens the need for manual labor.

TABLE 7.1
Composted Contents and Percentages

S. No	Contents	Percentage (%)
1	Organic material	81.01
2	Carbon (C)	30.72
3	Ash	30.715
4	Nitrogen (N)	1.39

The outputs are given in Figure 7.3. Figure 7.4 represents the structure of the proposed bin structure. By lowering the size necessary for categorizing and decomposing biodegradable waste, the special composting of waste based on the Berkeley rapid composting technique achieves better results than the standard method. Table 7.2 lists the components as referenced in Figure 7.4, and Table 7.3 shows the different components of the system along with their specifications.

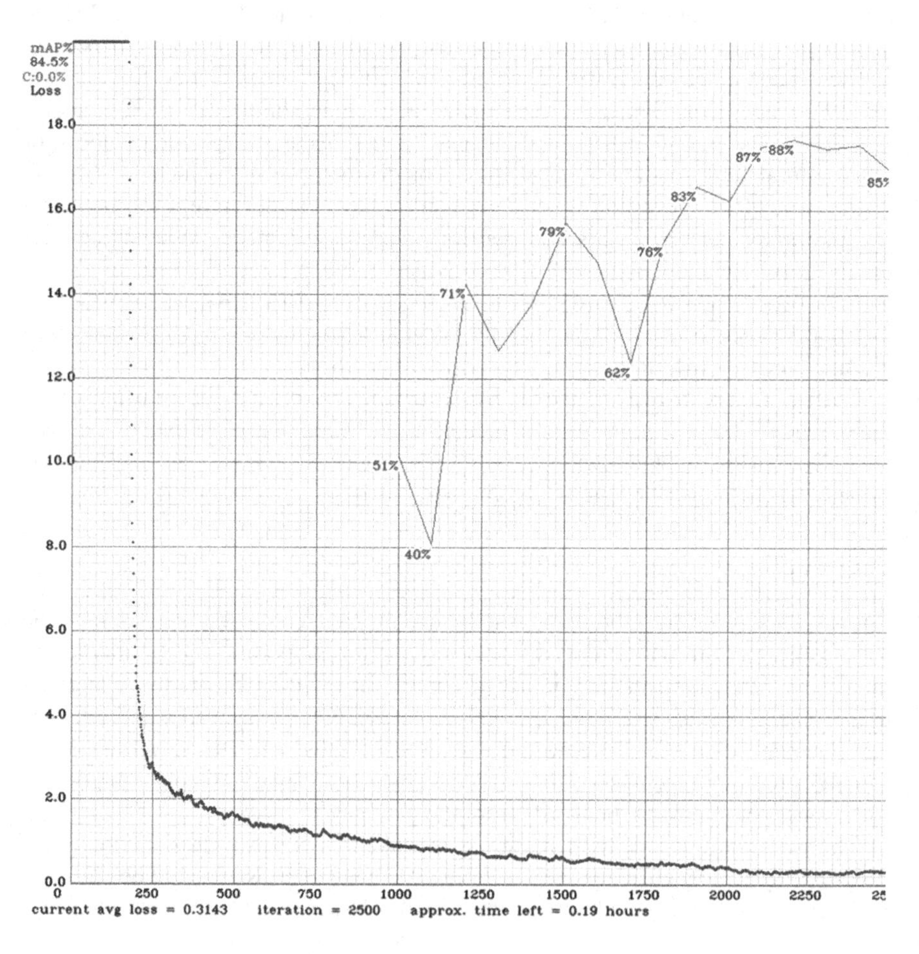

FIGURE 7.2 Mean average precision for 2500 epoch.

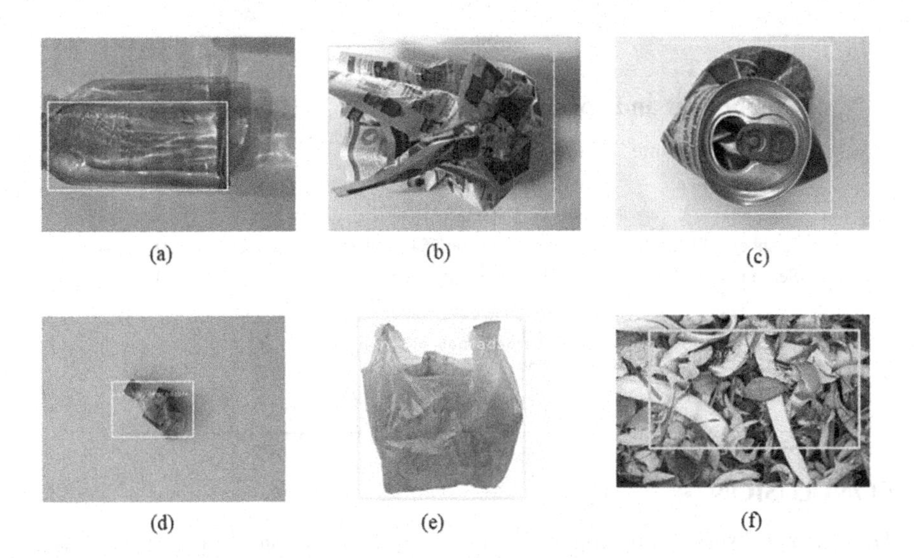

FIGURE 7.3 Model detection of biodegradable (b,f) and non-biodegradable (a,c,d,e) waste classified by the trained model.

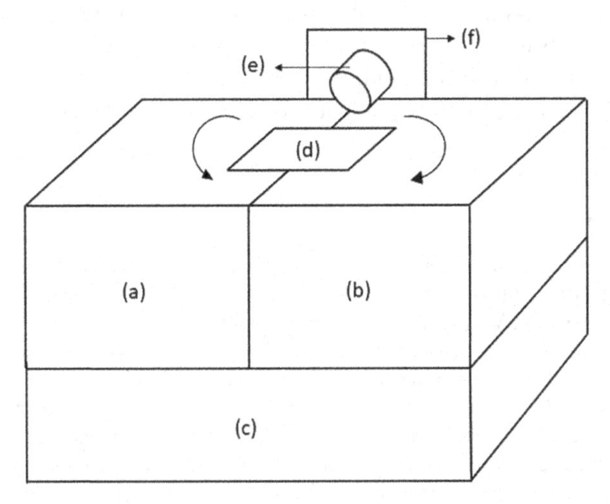

FIGURE 7.4 Structure of the proposed waste classifier and biocomposting system.

TABLE 7.2
Component as Referenced in Figure 7.4

Notations	Component
a	Biodegradable compartment
b	Non-biodegradable compartment
c	Biocomposting compartment
d	Waste in and flipping separator
e	Camera module
f	Processing module enclosing camera module

TABLE 7.3
Component and Specifications

Component Name	Specification	Number
Raspberry Pi 4	4GB RAM – 64bit, Cortex-A72Quad-core SoC @ 1.5GHz	1
Raspberry Pi camera	5MP Omnivision 5647 camera module	1
Servo motor	+5V typically – operating voltage T: 2.5kg/cm Speed:0.1 s/60° Rotation degree: 0°–180°	1
Battery	12V, 5Ah power safe battery	1

CONCLUSION

This chapter discusses the application of waste classification in the health industry with the help of AI. The proposal used machine learning and deep learning to solve the classification problem. The YOLO algorithm is utilized to classify the waste fed into the bin and SVMs were used to preprocess the image dataset used to train the algorithm. A waste composter based on Berkeley rapid composting was also attached to the unit. The works highlighted here primarily focus on waste classification to remove the mixture of waste that causes harm to both the environment and the ecosystem.

REFERENCES

1. Hoornweg, D. and Bhada-Tata, P. (2012). "A Global Review of Solid Waste Management." 1–116. https://documents1.worldbank.org/curated/en/302341468126264791/pdf/68135-REVISED-What-a-Waste-2012-Final-updated.pdf
2. Niehs.nih.gov. (2018). "Cancer and the Environment." 46. [Online]. Available: https://www.health.state.mn.us/communities/environment/hazardous/topics/cancerenvt.html#:~:text=Chemicals%20in%20the%20environment&text=Benzene%2C%20asbestos%2C%20vinyl%20chloride%2C,people%20are%20exposed%20to%20them.
3. Capel, C. (2008). "Waste sorting - a look at the separation and sorting techniques in todays European market." Waste Management World.
4. Belongie, S., Malik, J. and Puzicha, J. (2002). Shape Matching and Object Recognition Using Shape Contexts. IEEE Transactions on Pattern Analysis and Machine Intelligence, 24(4), 509–522.
5. Dror, R., Adelson, E. H. and Willsky, A. S. (2001). Recognition of Surface Reflectance Properties from a Single Image Under Unknown Real-World Illumination. In IEEE Workshop on Identifying Objects across Variation in Lighting, IEEE.
6. Liu, C., Sharan, L., Adelson, E. H. and Rosenholtz, R. (2010). Exploring Features in a Bayesian Framework for Material Recognition. In 2010 IEEE Conference on Computer Vision and Pattern Recognition (CVPR), IEEE, pp. 239–246.
7. Suzuki, J. (2020). Support Vector Machine. In: Statistical Learning with Math and R. Springer, Singapore, doi: 10.1007/978-981-15-7568-6_9.
8. Berk, R. A. (2020). Support Vector Machines. In: Statistical Learning from a Regression Perspective. Springer Texts in Statistics. Springer, Cham, doi: 10.1007/978-3-030-40189-4_7.

9. Redmon, J. and Farhadi, A. (2018). YOLOv3: An Incremental Improvement. arXiv preprint, arXiv:1804.02767.
10. Bui, H., Lech, M., Cheng, E., Neville, K. and Burnett, I. (2017). Object Recognition Using Deep Convolutional Features Transformed by a Recursive Network Structure. IEEE Access, 4, 10059–10066. doi: 10.1109/ACCESS.2016.2639543.
11. Mikołajczyk, A. and Grochowski, M. (2018). Data Augmentation for Improving Deep Learning in Image Classification Problem. 2018 International Interdisciplinary PhD Workshop (IIPhDW), Swinoujście, pp. 117–122. doi: 10.1109/IIPHDW.2018.8388338.
12. Azis, F., Suhaimi, H. and Abas, E. (2020). Waste Classification using Convolutional Neural Network. ITCC 2020: Proceedings of the 2020 2nd International Conference on Information Technology and Computer Communications, pp. 9–13. doi: 10.1145/3417473.3417474.
13. Ruiz, V., Sánchez, Á., Vélez, J.F. and Raducanu, B. (2019). Automatic Image-Based Waste Classification. In: Ferrández Vicente, J., Álvarez-Sánchez, J., de la Paz López, F., Toledo Moreo, J., Adeli, H. (eds) From Bioinspired Systems and Biomedical Applications to Machine Learning. IWINAC 2019. Lecture Notes in Computer Science, vol 11487. Springer, Cham. doi: 10.1007/978-3-030-19651-6_41.
14. Ahmad, K., Khan, K. and Al-Fuqaha, A. (2020). Intelligent Fusion of Deep Features for Improved Waste Classification. IEEE Access, 8, 96495–96504. doi: 10.1109/ACCESS.2020.2995681.
15. Dalos, D.E. (1997). Method for Composting Solid Waste. Journal of Cleaner Production, 5(3), 230. doi: 10.1016/S0959-6526(97)85775-7.
16. Khakimov, A., Zuparov, M. and Gulmurodov, R. (2020). Short Time Method of Composting for Cultivation of Button Mushroom [Agaricus Bisporus (Lange) Imbach]. Research on Crops, 21, 106–112.
17. Singh, S. and Nain, L. (2014). Microorganisms in the Conversion of Agricultural Wastes to Compost. Proceedings of the Indian National Science Academy, 80, 473–481. doi: 10.16943/ptinsa/2014/v80i2/4.
18. Raut, M., William, SPM, Bhattacharyya, J, Chakrabarti, T. and Devotta, S. (2008). Microbial Dynamics and Enzyme Activities During Rapid Composting of Municipal Solid Waste - A Compost Maturity Analysis Perspective. Bioresource Technology, 99, 6512–6519. doi: 10.1016/j.biortech.2007.11.030.
19. Mccartney, D. and Tingley, J. (2013). Development of a Rapid Moisture Content Method For Compost Materials. Compost Science & Utilization, 6, 14–25. doi: 10.1080/1065657X.1998.10701927.
20. Behera, S. K., SR Y, Aouthithiye Barathwaj, L, Vasundhara, G, Saisudha, N C, Haariharan and J, Sathya Priya. (2020). AI Based Waste Classifier with Thermo-Rapid Composting. In 2020 International Conference on Power, Energy, Control and Transmission Systems (ICPECTS), Chennai, India, pp. 1–4. doi: 10.1109/ICPECTS49113.2020.9337012.

8 SmartWear
An IoT-Based Integration of Home Automation and Healthcare Watch

Saswati Kumari Behera, Saisudha G, and Vasundhara L
Sri Sai Ram Engineering College, Chennai, Tamil Nadu, India

INTRODUCTION

Everything is smarter now that we are in a new generation. One of the trendiest emerging technologies is wearable technology, which is anticipated to keep expanding due to strong demand. The practice of creating consumer apps is expanding alongside the field. When it comes to personalizing and securing users so they may be easily controlled and relocated, device health and control are very important. Today's world is smaller and more technologically advanced. To keep up with our lives and make them easier, we all rely on one another. By automating everything and continually monitoring and updating individuals, technology is essential in assisting people in simplifying their life.

Using the Internet, it is possible to automatically and instantly control equipment from anywhere in the world. A network of devices connected to the Internet via different protocols, including Bluetooth, Wi-Fi, LoRaWAN, etc., is used for home automation. Sensors built into many of these Internet of Things (IoT) devices monitor changes in light, temperature, humidity, and motion. This allows consumers to find out more about the institution where the gadget is located. Actuators are genuine mechanical parts, such as motors, valves, or switches, that allow users to remotely operate a device and physically alter it [1–3].

Traditional medicine based on biotechnology is becoming digitized thanks to advancements in technology and scientific theory; you can instantaneously and automatically control equipment using the Internet from any location in the world. A network of Internet-connected devices and a variety of protocols, including LoRaWAN, Wi-Fi, Bluetooth, ZigBee, etc., are used for home automation. Many of these IoT devices are equipped with sensors that detect changes in light, temperature, humidity, and motion. Users can learn more about the gadget's setting thanks to this. Genuine mechanical components known as actuators, such as motors, valves, or switches, let users control a device remotely and physically affect it. Because of improvements in

DOI: 10.1201/9781003388319-8

technology and scientific theory, traditional medicine based on biotechnology has gradually begun to be digitalized.

RESEARCH METHODOLOGY

Medical Care

Our monthly schedule definitely includes healthcare. Health check-ups are conducted at clinics and hospitals to monitor patients' or healthy individuals' health statuses and assure their well-being. Blood pressure, electrocardiograms, and blood testing are often employed procedures. Only clinics and hospitals are equipped with these techniques. Using wearable technologies, such as fitness trackers and other wirelessly connected devices, patient therapy can be tailored to each patient. These devices can be configured to send appointment reminders, changes in blood pressure, calorie intake, activity monitoring, and a wide range of other events [4–6]. As the technologies like IoT (Internet of Things) provides significant advantages as illustrated in Figure 8.1, it greatly benefitted elderly patients by constantly checking their health status. The most affected are families and single people.

When the user's typical actions are modified or interfered with, the alarm system alerts the appropriate healthcare providers and family members. Doctors can more effectively keep an eye on the health of their patients by implementing IoT-enabled wearables and other home monitoring technology. They can monitor how closely patients are adhering to their treatment regime and spot illnesses that call for quick attention. Doctors can select the most efficient treatment plan for patients and achieve the intended result with the help of data from IoT devices [7–9]. The significance of point-of-care technologies and their capacity to give efficient solutions to several unmet healthcare demands continue to be highlighted as there is growing focus paid to healthcare-related issues on a global scale. Furthermore, the need for research and development of effective and affordable healthcare delivery techniques has been sharply underlined by the current crisis in healthcare spending.

If the user's normal activities are altered or disrupted, the alert system notifies appropriate medical personnel and family members. IoT-enabled wearables and other home monitoring tools can help doctors track patient health more effectively. Using data from IoT devices, healthcare professionals can choose the most effective treatment for the patient and achieve the desired results [7–9]. At a time when

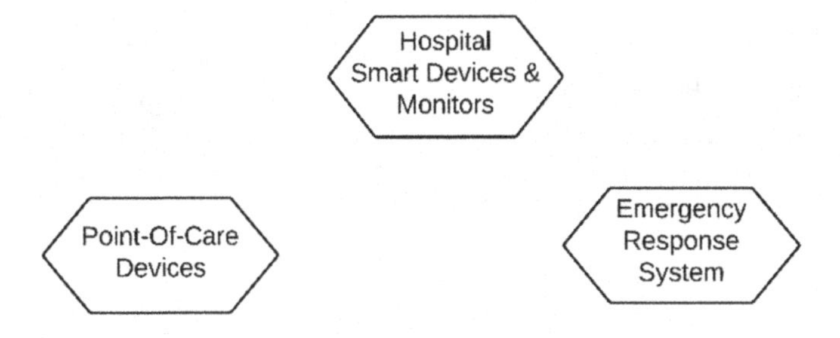

FIGURE 8.1 The advantages of IoT-based medical care.

healthcare issues are increasing across the globe, the importance of point-of-care technology and its ability to provide effective solutions to various unmet healthcare needs continues to be highlighted.

AUTOMATION

The main purpose of an automated system is to simplify automation. Automating home appliances online using IoT will make things smarter. Devices are now controlled by home automation [10]. A mobile app-controlled home automation system allowing users to use their phones to perform specific tasks. "Home automation" is the term used to describe the automated, computerized management of household tasks, functions, and appliances. In short, this means that you can easily control functions and devices in your home online to improve comfort, improve security, improve production & operation and even reduce costs & manual errors which is depicted in Figure 8.2.

The term home automation refers to the mechanical management of smart technological devices in the home. Home automation systems organize pre-programmed smart devices to meet the homeowner's needs. Where home automation gets really "smart" is in the networked devices that connect to and manage the network [11, 12].

No direct wiring is required when using a radiofrequency (RF)-based wireless remote control system to turn devices on and off from anywhere in the home. Automate your devices faster and more efficiently with voice assistants like Alexa, Google Assistant, and more.

Home appliances can be controlled and operated with voice commands. Users benefit from Google Assistant as it can only be used with voice commands for healthcare as well. Healthcare workers can closely monitor the patient using wearable devices, remote monitoring devices, and smart home technology [13–17]. Home automation systems allow devices to be activated independently of each other. Users can also schedule automated processes such as temperature changes, lighting changes, entertainment system calibration, and more. What is home automation? It is a comprehensive strategy that can make everyday life easier and reduce certain costs.

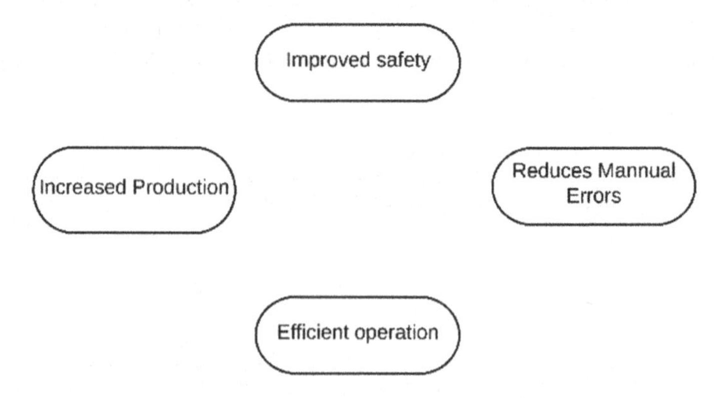

FIGURE 8.2 Advantages of automation in the industrial sector.

Cloud Computing

One of the many important developments in technology over the past 10 years is cloud computing. The use of remote computers connected through the Internet can be used to store, manage, and process data. Before cloud computing was created, the majority of users had to store their data on real hard drives, which were expensive and regularly deteriorated over time. The majority of organizations invested a significant sum of money in the maintenance, management, and growth of their data centers. As a result, the concept of cloud computing was created to take the place of the old technology.

Cloud computing has many advantages for the information technology (IT) sector, including lower costs and more effectiveness. Cloud computing is particularly versatile because it leverages the paper model and has a low barrier to entry. A cloud provider can supply several resources to a range of clients in addition to assisting with infrastructure management. As a result, companies who rely on third parties could concentrate on other objectives like expanding their clientele and product line. The greatest impact of cloud computing on IT professions is projected to occur within the next 5–10 years.

Both corporations and individual clients benefit from cloud-based services. From a global viewpoint, it is evident that companies of all sizes fully utilize cloud computing to create and shape their business models, allowing them to concentrate on other projects while growing their customer and client bases. For people who wish to work in IT in the future, being current on technology is essential. Being open-minded and eager to learn are equally important. Due to the rapid advancement of technology, those who do not continue their education risk falling behind.

IoT (Internet of Things)

IoT is the network of connected computers and devices that can share data without involving any human or gadget intervention.

People with implanted heart monitors, automobiles with built-in tire pressure monitors, and other scenarios are some of the things that can receive Internet Protocol (IP) addresses and transmit data over a network. Businesses across all industries are rapidly using the IoT to increase productivity, better understand customers, and provide better customer service. The IoT ecosystem consists of smart devices that use embedded systems, such as various processors, analog and digital sensors, and communication devices, to collect, transmit, and respond to information they learn about their environment. IoT gateways and other edge devices connect to IoT devices and can analyze data locally or send it to the cloud for analysis. These devices regularly communicate with other similar devices and respond to the data they exchange. Users can use the device to set preferences, provide instructions, or get information, but it does most of the work itself. The specific IoT application used strongly affects the connectivity, networking, and communication protocols of these network-enabled devices. The IoT can also use machine learning and artificial intelligence (AI) to simplify and improve the data collection process.

Medical Care with Automation

Home automation has a positive impact on healthcare as well. Smart health monitoring systems use pulse sensors to measure a person's blood pressure and pulse.

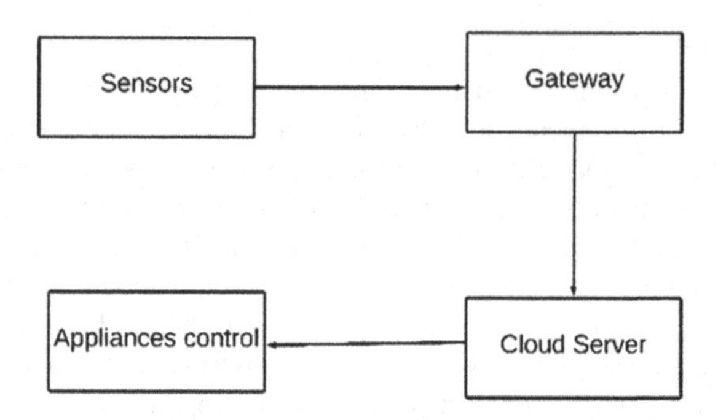

FIGURE 8.3 The architecture of medical care and automation.

The home automation system controls devices based on user commands. When used together, the two technologies work as a seamless system. To protect patient health, the device is automated, with increased ventilation and automatic messages sent to designated caregivers when the patient's pulse and blood pressure are lower or higher.

Merging these two systems will help patients in emergencies [18]. E-health is a hot topic of AI in marketing, healthcare, and other related fields. It refers to medical information and services enhanced or provided through the Internet and similar technologies. In smart homes that incorporate wearable medical devices, environmental sensors, actuators, and the latest communication and information technologies, the health and well-being of older people can be monitored continuously and remotely. Thanks to smart homes, older people no longer need to pay for expensive and substandard medical facilities and can live in the comfort of their own homes.

Medical experts can also monitor the general health of seniors, and they can provide advice and support via the Internet. Figure 8.3 illustrates a simple architecture of medical care and automation.

Wearable Technology

The inevitable combination of science, technology, and health has enabled the widespread adoption of the Health Network and will transform medicine. Wearable medical and health devices can collect real-time user health data, including user behavior, mood, and sleep. This knowledge is extremely valuable in social and business contexts. In healthcare, wearables serve as important network nodes, connecting hospitals and patients with sensors and IoT technologies. Thus, a vast medical network has been created. Portable devices, when uploading data to the cloud, are particularly vulnerable to attacks and data breaches due to their ability to collect user data anytime, anywhere. Technical flaws make it more likely that information and privacy leaks will occur. If the portable device cannot restrict data flow connections. Currently, how to ensure information security and user privacy while collecting data with wearable healthcare devices is an area that needs further research. This chapter examines medical wearables data from a technical, administrative, and legal perspective to protect user privacy and data security while promoting long-term growth and scientific data collection for the medical wearable industry. The importance of health in everyday life cannot be overemphasized. As the saying goes, "the body is the capital of the revolution." With the explosive growth of China's economy, the standard of living has changed dramatically in recent years, and the aging population has raised public concern about health issues. In particular, overall public health is horrendous, with urban populations having long-term diet patterns and chronic diseases. In this regard, smart wearable medical products are due to the growth of smart medical services. These technologies combine IT with healthcare providers, networks, big data, and cloud computing to deliver personalized healthcare to customers.

Wearable electronics is not a new trend. Over the past decades there have been countless attempts to design and build a variety of wearable devices that can be used for various activities. In the 2010s, fitness-related features dominated wearable tech products. However, interest in wearable technology has increased significantly in recent years. The new devices that emerged were more sophisticated than their predecessors, used for a variety of tasks, and had the potential to improve performance in certain occupations. Explore the opportunities and challenges of adopting wearable technology in the healthcare sector within the framework of disruptive innovation theory. This tactic is beneficial because it clarifies the process and its implications

EXPERIMENT

The system's purpose is to deliver to consumers a convenient way to automate their machines and monitor their health on wearable devices. The pulse sensor, the GPS module, and the ESP NodeMCU module constitute the proposed SmartWear system. For home automation and health monitoring, SmartWear leverages cloud computing and the IoT. Data analysis using Adafruit's cloud platform. Google Assistant can be used to remotely control devices in your home.

SmartWear has three main functions

1. Health monitoring
2. Home automation
3. Integration of home automation and health monitoring

Figure 8.4 shows the process used by the SmartWear fitness tracking system. The main elements of SmartWear include a heart rate monitor, a sensor, and a pulse sensor. After it is connected, the simplicity of flexible app development provides a foundation that encourages convenience. The ESP NodeMCU transmitter can send

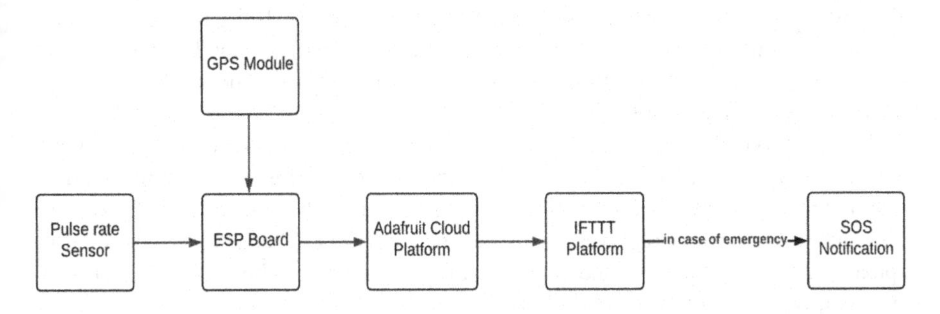

FIGURE 8.4 SmartWear's health monitoring system.

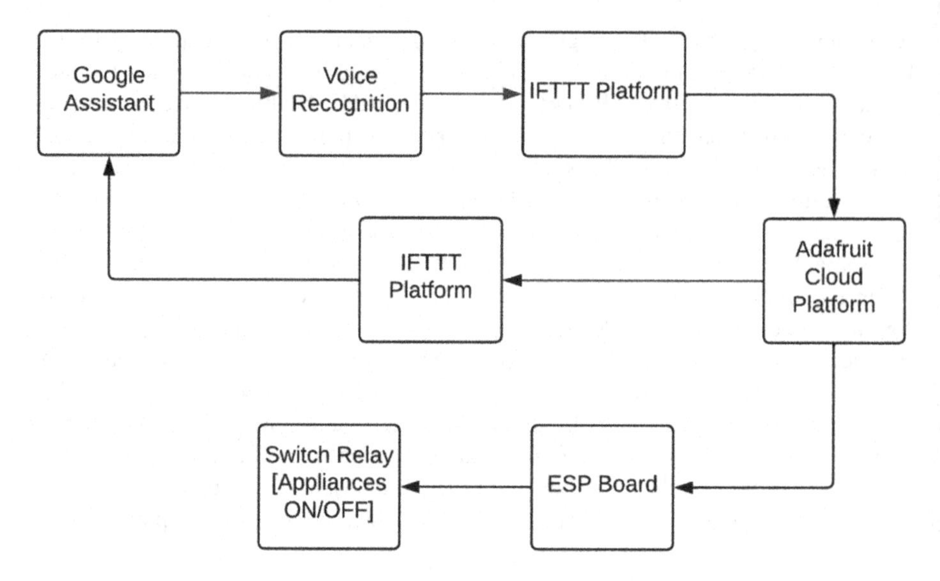

FIGURE 8.5 SmartWear's home automation system.

the information obtained from the pulse sensor to the connected mobile phone. Pulse data is transmitted from the sensor to the phone while connected, where it is processed in the app and published to Adafruit's cloud platform.

Information about blood flow is sent to the ESP through a pulse sensor. Each heartbeat causes the sensor LED to flash in time.

The Adafruit IO platform receives the calculated pulses, where the data is represented in a counter. An alarm email is sent to the user's contacts via IFTTT when the heart rate exceeds a preset threshold, coupled with the user's location as determined by the GPS. Most mid to high-end watches come with GPS locating by default, thus smart watches can simply utilize this service.

While the user is streaming after a fall or conducting outdoor monitoring, the installed GPS is employed as a positioning system. If this functionality is supported, GPS will be able to direct family members or emergency personnel where to look and assist users who have fallen.

Figure 8.5 illustrates the operation of the SmartWear home automation system. To control "the device," users first enter commands into the watch via Google Assistant. The Adafruit cloud platform receives commands from the IFTTT platform, processes them, and sends them to the ESP board. The Adafruit cloud interface simplifies the management of multiple devices and sends return commands to Google Assistant via an IFTTT link.

The method of integrating home automation and health monitoring with SmartWear is shown in Figure 8.6. The wrist sensor of this SmartWear function detects the user's wrist and transmits the information to the controller at the watch. The onboard controller receives data from the external ESP controller. The data is uploaded to the Adafruit cloud platform after processing. The widget-based data presentation is hosted by a cloud platform.

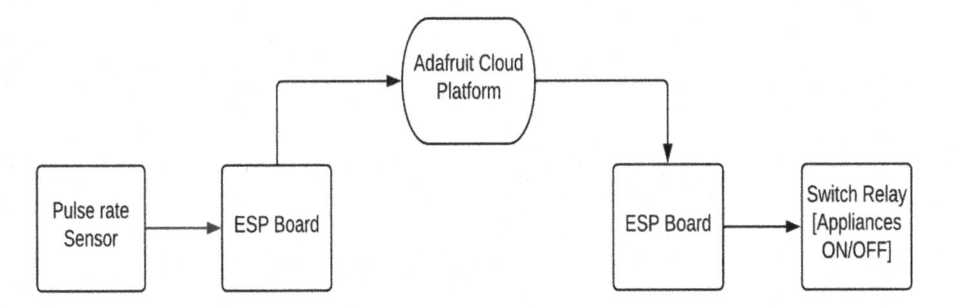

FIGURE 8.6 The block diagram illustrates the integration of both systems.

FEED VALUE (BPM)	APPLIANCES (ON/OFF)
93	OFF
116	OFF
127	ON
63	ON
102	OFF
164	ON
97	ON

FIGURE 8.7 Feed values obtained in the cloud and the relative control of appliances.

The relays are used to manage the load of the card according to the data received by the ESP.

Figure 8.6 illustrates how SmartWear will integrate with home automation and health monitoring. For this SmartWear function, a pulse sensor measures the user's pulse and sends the data to the watch's microcontroller. Data are sent from the external ESP to the board's internal ESP, preparing to upload information to Adafruit's cloud. On the cloud, the data is displayed. The diagram in Figure 8.6 shows the integration of SmartWear with home automation and health monitoring. A pulse sensor monitors the user's pulse and sends the information to the watch's ESP for this SmartWear feature. The board's internal ESP receives data from the external ESP. The information is cleaned before being uploaded to the Adafruit cloud computing platform. Widgets are used to display data on cloud platforms. Figure 8.7 shows the different feed values obtained in the cloud and the relative control of appliances.

RESULTS

The relays are used to regulate the charge of the card according to the information received by the ESP. The proposed SmartWear system is already tested. The position of the user is transmitted using a GPS module and the sensor gathers data and is stored on a cloud server (Figures 8.8–8.10).

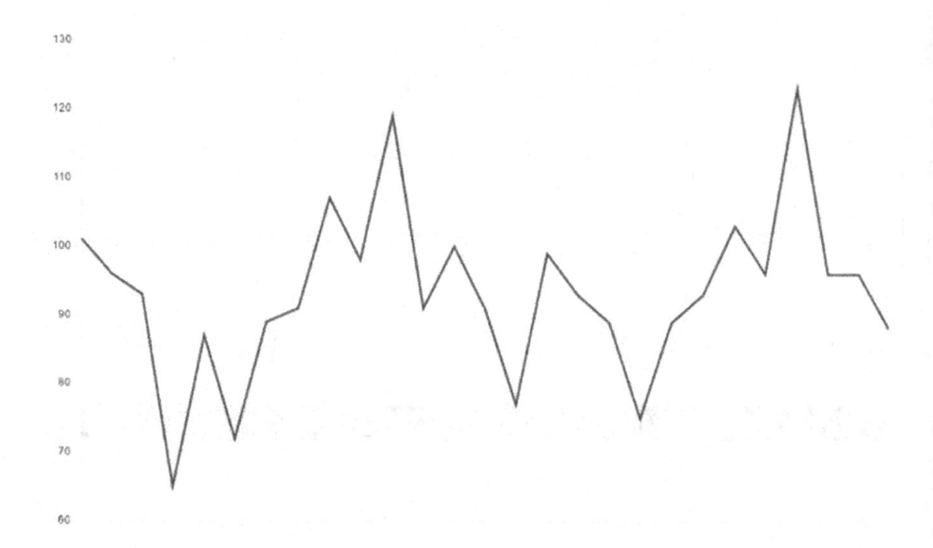

FIGURE 8.8 The user's pulse rate in the cloud.

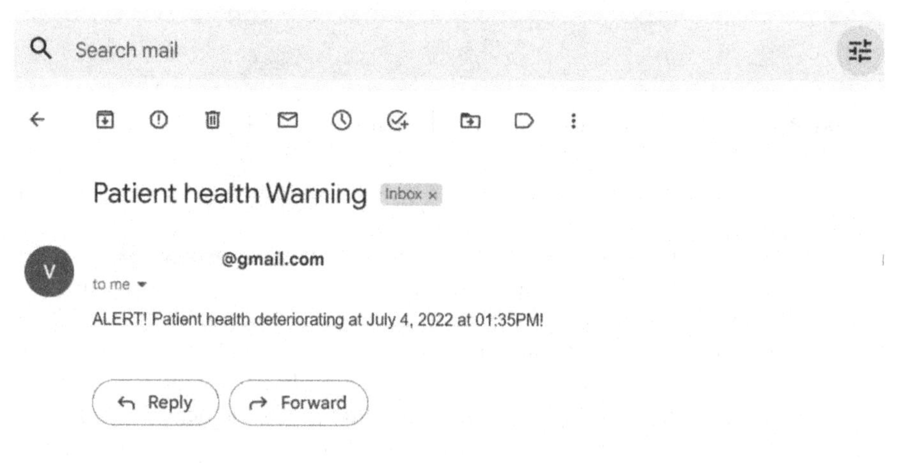

FIGURE 8.9 Emergency email alert.

CONCLUSION

The suggested solution combines home automation and health monitoring, enabling users to automate care for their convenience without further dependencies. Users of SmartWear can automate their devices and keep an eye on their health from the comfort of their own homes. It enables users to regularly monitor their heart rate and pulse. Home automation services exhibit their potential to deliver an aesthetic advantage in today's unpredictable climate by utilizing the most recent technologies. Notwithstanding all of the challenges, we must keep implementing numerous services for societal benefit to make the world smarter.

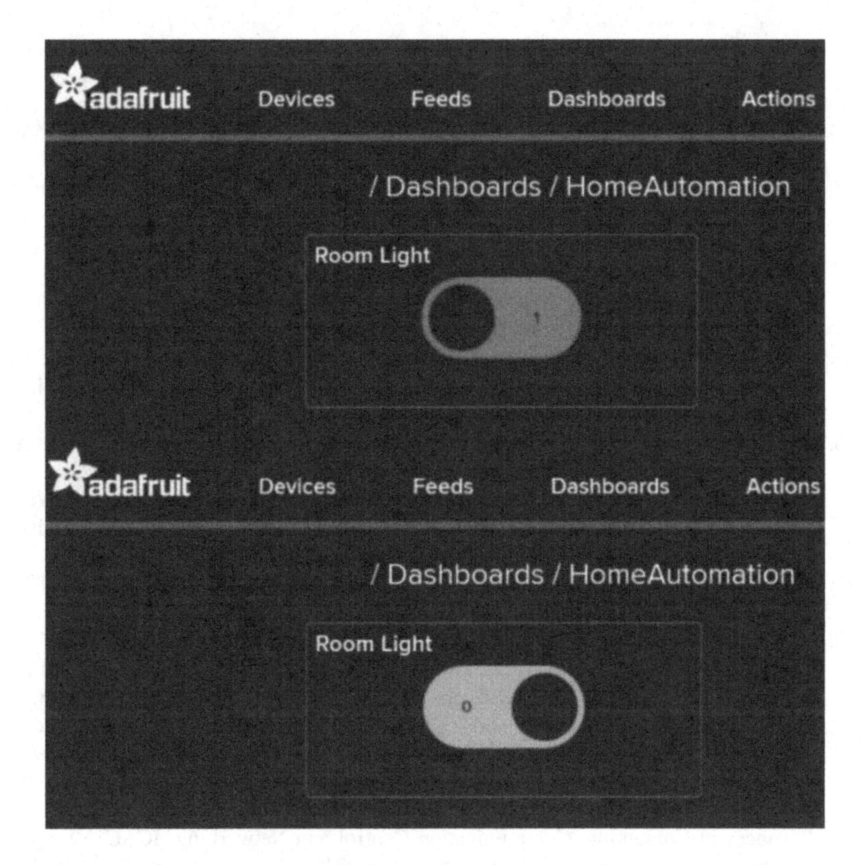

FIGURE 8.10 Adafruit cloud dashboard.

REFERENCES

1. M. Al-Kuwari, A. Ramadan, Y. Ismael, L. Al-Sughair, A. Gastli and M. Benammar, "Smart-Home Automation Using IoT-Based Sensing and Monitoring Platform," 2018 IEEE 12th International Conference on Compatibility, Power Electronics and Power Engineering (CPE POWERENG 2018), 2018, pp. 1–6, doi:10.1109/CPE.2018.8372548.
2. A. Doshi, Y. Rai and D. Vakharia, "IoT Based Home Automation," 2021, doi:10.22214/ijraset.2021.37287.
3. M. Asadullah and A. Raza, "An Overview of Home Automation Systems," 2016 2nd International Conference on Robotics and Artificial Intelligence (ICRAI), 2016, pp. 27–31, doi:10.1109/ICRAI.2016.7791223.
4. A. Kumar, R. Krishnamurthi, A. Nayyar, K. Sharma, V. Grover and E. Hossain, "A Novel Smart Healthcare Design, Simulation, and Implementation Using Healthcare 4.0 Processes," in IEEE Access, vol. 8, pp. 118433–118471, doi:10.1109/ACCESS.2020.3004790.
5. T. M. Ghazal, M. K. Hasan, M. T. Alshurideh, H. M. Alzoubi, M. Ahmad, S. S. Akbar, B. A. Kurdi and I. A. Akour, "IoT for Smart Cities: Machine Learning Approaches in Smart Healthcare—A Review," in Future Internet, vol. 13, no. 8, p. 218, 2021, doi:10.3390/fi13080218.

6. Alansari, Z., Soomro, S., Belgaum, M. R. and Shamshirband, S. "The Rise of Internet of Things (IoT) in Big Healthcare Data: Review and Open Research Issues," in Saeed, K., Chaki, N., Pati, B., Bakshi, S. and Mohapatra, D. (eds) Progress in Advanced Computing and Intelligent Engineering. Advances in Intelligent Systems and Computing, vol 564. Springer, Singapore, doi: 10.1007/978-981-10-6875-1_66.
7. Verma, G., Prakash, S. (2021). "Internet of Things for Healthcare: Research Challenges and Future Prospects," in Hura, G.S., Singh, A.K., Siong Hoe, L. (eds) Advances in Communication and Computational Technology. ICACCT 2019. Lecture Notes in Electrical Engineering, vol 668. Springer, Singapore, doi: 10.1007/978-981-15-5341-7_80.
8. P. Valsalan, A. Tariq and A. Hussain, "Iot Based Health Monitoring System," in Journal of Critical Reviews, vol. 7, no. 4, pp. 739–742, 2020, doi:10.31838/jcr.07.04.137.
9. S. Baker, W. Xiang and I. Atkinson, "Internet of Things for Smart Healthcare: Technologies, Challenges, and Opportunities," in IEEE Access, pp. 1–1, 2017, doi:10.1109/ACCESS.2017.2775180.
10. K. K. Tan, T. H. Lee and C. Y. Soh, "Internet Based Monitoring of Distributed Control Systems," in Energy and Power Engineering, vol. 45, no. 2, pp. 128–134, 2002.
11. E. Johri, "Remote Controlled Home Automation Using Android Application via Wi Fi Connectivity," in International Journal on Recent and Innovation and Recent Trends in Computing and Communication, vol. 3, no. 3, pp. 2321–8169, 2012.
12. K. SM, "Assistant Professor, Department of telecommunication, MSRIT, Bangalore, India," "Home Automation Using Internet of Things" July 1999.
13. Y. Mittal, P. Toshniwal, S. Sharma, D. Singhal, R. Gupta and V. K. Mittal, "A Voice Controlled Multi-Functional Smart Home Automation System," 2015 Annual IEEE India Conference (INDICON), 2015, pp. 1–6, doi:10.1109/INDICON.2015.7443538.
14. M. A. Chahinez, Z. Hamaizia and Zahra, "Internet of Things (IoT) Based Home Automation System Using Google Assistant," 2020, doi:10.13140/RG.2.2.17650.73926.
15. S. S. Yadav, P. Kumar, S. Kumar and S. Singh, "Google Assistant Controlled Home Automation with Voice Recognition," 2021 3rd International Conference on Advances in Computing, Communication Control and Networking (ICAC3N), 2021, pp. 1039–1044, doi:10.1109/ICAC3N53548.2021.9725502.
16. M. P. Gupta, Department of Electronics and Communication, Maharishi Dayanand University, Rohtak, Haryana, India, "Google Assistant Controlled Home Automation" Volume: 05 Issue: 05, May-2018, https://ieee-vecsb.org/wp-content/uploads/sites/45/Google-assistant-controlled-home-automation.pdf
17. A. Agarwal, A. Sharma, A. S. Samad and S. Babeetha, "UJALA Home Automation System Using Google Assistant" Volume: 04 Issue: Feb-2018, https://ijariie.com/AdminUploadPdf/Home_Automation_Using_Google_Assistant_ijariie8241.pdf
18. P. N. Dawadi, D. J. Cook and M. Schmitter-Edgecombe, "Automated Cognitive Health Assessment Using Smart Home Monitoring of Complex Tasks," in IEEE Transactions on Systems, Man, and Cybernetics: Systems, vol. 43, no. 6, pp. 1302–1313, 2013, doi: 10.1109/TSMC.2013.2252338.

9 An Analytical Comparison of the Identification of Non-Small Cell Lung Cancer Nodules Using CT Scans and Prominent Deep Learning Models

Sunil Kumar and Vishal Awasthi
School of Engineering and Technology (UIET),
CSJM University, Kanpur, UP, India

Amar Pal Yadav
Noida Institute of Engineering and
Technology, Greater Noida, UP, India

Shivneet Tripathi
School of Engineering and Technology (UIET),
CSJM University, Kanpur, UP, India

Prachi Chhabra
JSS Academy of Technical Education, Noida, UP, India

INTRODUCTION

Non-small cell lung cancer (also known as NSCLC) is responsible for the majority of lung cancer diagnoses worldwide. Lung cancer is one of the most commonly seen types of the disease worldwide. In the world of cancer, NSCLC is the leading cause of mortality. The NSCLC subtype of lung cancer is responsible for roughly 85% of all instances of lung cancer [1]. NSCLC is often diagnosed in advanced stages, resulting in poor patient outcomes. Early detection of NSCLC is critical for effective treatment and improved survival rates as well as crucial for effective treatment and a better prognosis. It is called "non-small cell" because the cancer cells are larger and

DOI: 10.1201/9781003388319-9

more visible than the small cells that are discovered. NSCLC can grow and spread more slowly than small-cell lung cancer, and it is typically treated differently [2].

There are three main types of NSCLC, each of which is classified based on the way the cancer cells look under a microscope:

- Adenocarcinoma: This is the most prevalent form of NSCLC, accounting for around 40% of cases. It is common for adenocarcinoma to begin in the peripheral regions of the lungs, but it may eventually progress to other areas of the body.
- Squamous cell carcinoma: This subtype of NSCLC accounts for around 25–30% of all cases. It is common for squamous cell carcinoma to begin in the central regions of the lungs, such as the bronchi, but the disease may also spread out to other areas of the body.
- Large cell carcinoma: This is a less common type of NSCLC, accounting for about 10–15% of cases. Large cell carcinoma can start anywhere in the lung and can grow and spread quickly.

The symptoms of NSCLC include a chronic cough, chest discomfort, shortness of breath, wheeze, exhaustion, and weight loss. NSCLC treatment choices vary on the cancer's stages, kind, and the patient's general condition. Due to the intricacy of the illness and the diversity of emotions that might be experienced, detecting emotions in lung cancer patients is a difficult process. NSCLC can develop in any part of the lung and is usually associated with long-term exposure to cigarette smoke, although it can also occur in non-smokers. Other risk factors for NSCLC include exposure to air pollution, radiation, or occupational hazards such as asbestos. Table 9.1 presents lung nodule sizes with their description [3].

Imaging tests like x-rays and computed tomography (CT) scans, as well as biopsies, which involve the removal of a tiny tissue sample for the purpose of evaluation under a microscope, are often used in conjunction with one another to arrive at a diagnosis. CT scan is a commonly used imaging modality for NSCLC detection. X-rays are used in CT scans in order to obtain detailed images of the interior of the body. CT scans are commonly used for NSCLC detection and are effective at identifying lung nodules that may be cancerous. They can detect nodules as small as 1–2 mm, and from the images, one can establish the size of the nodules, their form, and where they are located [4].

TABLE 9.1
Lung Nodule Sizes [3]

Nodule Size	Description
<1 cm	Small tumor
1–3 cm	Medium-sized tumor
3–5 cm	Large tumor
5–7 cm	Very large tumor
>7 cm	Extremely large tumor

Computer-aided diagnosis (CAD) systems based on deep learning algorithms have shown promising results in detecting NSCLC. CT scans are commonly used for lung cancer detection, and recent advancements in artificial intelligence have enabled the development of machine learning (ML) and convolutional neural networks (CNNs) for automated diagnosis [5].

The purpose of this investigation is to assess how well four modern deep learning models perform compared with one another (VGG19 [6], ResNet50 [7], Inceptionv3 [8], and Densenet-121 [9]) for NSCLC detection. VGG19, ResNet50, InceptionV3, and DenseNet-121 are deep CNNs that have been used for image classification tasks, including the classification of lung cancer in medical imaging. In recent years, researchers have explored the use of VGG19, ResNet50, Inceptionv3, and Densenet-121 to detect NSCLC in medical images such as CT scans and chest x-rays.

The basic approach is to use the VGG19, ResNet50, Inceptionv3, and Densenet-121 models as a feature extractor to extract relevant features from the medical images that may indicate the presence of NSCLC. The model is typically pre-trained on large-scale image datasets and then fine-tuned on the medical image data to adapt it for the specific task of lung cancer detection. The fine-tuned VGG19, ResNet50, Inceptionv3, and Densenet-121 models can be used to classify new medical images as either containing or not containing NSCLC, and the output of the model can be used to estimate the likelihood of NSCLC in the image [6–9].

As with any deep learning model, it is important to note that VGG19, ResNet50, Inceptionv3, and DenseNet-121 should always be used as an aid to support clinical decision-making rather than a replacement for human expertise. Medical imaging should always be interpreted by trained medical professionals who can provide context and additional information that may not be captured by the model. Additionally, further research and validation are needed before these methods can be widely adopted in clinical settings.

BACKGROUND

In a variety of image classification tasks, including medical imaging, deep learning models have shown efficacy that is both promising and successful. VGG19, ResNet50, Inceptionv3, and Densenet-121 are some of the most popular deep -learning models. These models have been pre-trained on large datasets and can be fine-tuned on specific tasks such as NSCLC detection.

Numerous studies found that, initially, we take an input image, such as a CT scan. Then a dataset of these photos is preprocessed before lung segmentation. Lung segmentation involves separating the lung from the body architecture including ribs, arteries, and blood veins. Lung nodules are detected after segmentation. The required features are provided to the ML model after feature extraction. Lung health is classified using the ML model [5]. Ardila et al. developed a CNN that was able to detect lung cancer with a sensitivity of 94.4% and a specificity of 93.9%. The CNN was trained on a dataset of over 42,000 CT scans and was able to accurately identify lung nodules that were missed by radiologists [10]. Priyadharsini et al. looked for abnormalities in the lungs using low-dose CT (LDCT) scanned pictures in their research. These abnormalities might be cancer or could lead to the development of cancer. A median filter is used during the acquisition and preprocessing of a three-dimensional picture of the lungs obtained from an LDCT scan [11]. Khehrah et al. suggested automated lung segmentation. Histograms were used to estimate thresholds by constructing and evaluating picture histograms. Lungs isolated from the picture after the threshold-detected outside area are deleted. Lung segmentation included morphological and linked components analysis. Lung segmentation yields zones of interest; they demonstrated an automated method for identifying lung nodules from CT scan pictures. After lung segmentation, they identified nodules. Otsu was employed to find parenchyma nodules by differentiating blood vessels, bronchi, and nodule interior features [12]. Three interdependent deep-fusion learning algorithms for lung nodule identification from CT images were created. Hierarchical deep-fusion models include multi-perspective hierarchical deep-fusion learning mode (MPF), single feature & multi-perspective hierarchical deep-fusion (SFMPF), and single feature & multi-perspective hierarchical deep-fusion (MFMPF). The MPF model is categorized into three different levels by using multi-perspective deep fusion. The SFMPF model is an image-feature-based hierarchical deep fusion learning algorithm. The use of bilateral, trilateral, Gabor, and LoG-filtered imagery is put to the test to evaluate the performance of four unique feature image-based hierarchical deep-fusion learning models. Integrating these four SFMPF models results in the creation of the multi-feature multi-perspective hierarchical deep-fusion learning model [13]. After performing manual segmentation, CNN was used to analyze the CT image. Although the results of the segmentation based on DeepLabv3 and VGG19 are better than the results of the artificial segmentation, the testing revealed that both SegNet and the artificial segmentation outcomes almost overlap. During the same period, SegNet identified 120 individuals with benign lung nodules and 120 individuals with early lung cancer. Both diagnoses were made by pathological examination [14].

MATERIAL AND METHODS

Dataset

In this study, the lung cancer collection from the Iraq-Oncology Teaching Hospital/National Center for Cancer Diseases (IQ-OTH/NCCD) was utilized. It incorporates CT scans with lung cancer at varying stages, in addition to those of healthy participants. Oncologists and radiologists were the ones in charge of marking the slides for the IQ-OTH and NCCD. The collection is comprised of a total of 1067 images, each of which represents a CT scan slice taken from one of 110 patients. The dataset was created to facilitate the development of CAD and algorithms for lung cancer detection. It contains high-resolution CT scans of the chest, along with annotations of lung nodules provided by four experienced radiologists. The annotations include information such as the nodule location, size, shape, margin, and texture. The images were preprocessed to remove noise and normalize the intensity values. The instances of the CT scan with and without lung nodules are presented in Figure 9.1. These images were taken from a publicly accessible dataset [15].

Deep learning algorithms are a type of artificial neural network that can be trained to recognize patterns in data. VGG19, ResNet50, Inceptionv3, and Densenet-121 are

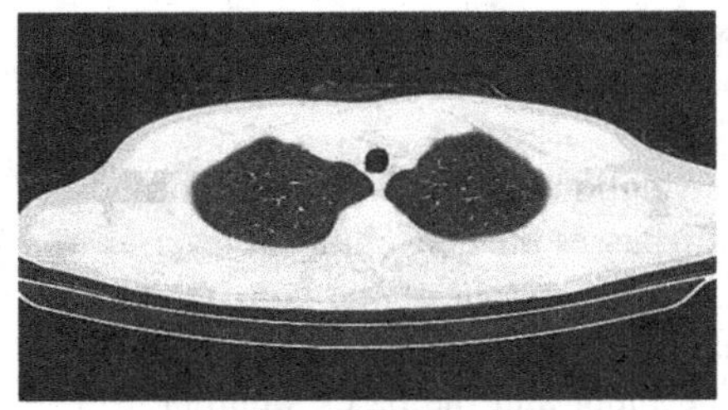

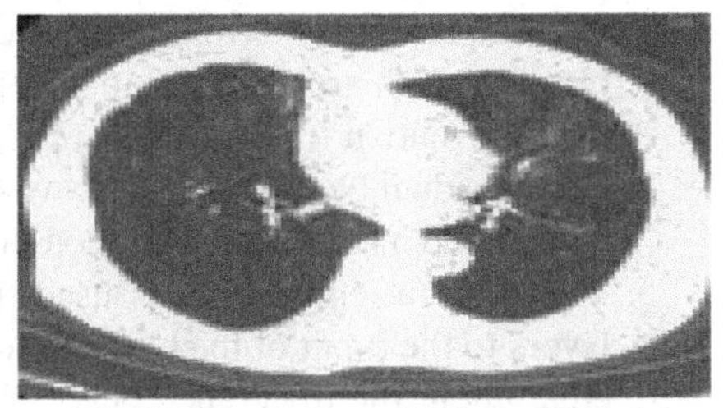

(A) Normal CT Scan Image

(B) CT Scan Image with Lung Nodule

FIGURE 9.1 Instances of CT scans [15].

all types of CNNs that have been trained on the dataset to identify suspicious nodules or masses in the lungs.

VGG19

VGG19 is a CNN that was introduced in 2014 by researchers at the Visual Geometry Group (VGG) at the University of Oxford. It is a deep learning model with 19 layers, comprising 16 convolutional layers and three fully connected layers. The main idea behind VGG19 is to use small 3 × 3 convolutional filters throughout the network, with max-pooling in between the convolutional layers to reduce the spatial dimension of the input. This results in a very deep network that can learn a wide range of features at different scales.

The input to the model is typically a 224 × 224 RGB image, which is passed through a series of convolutional and pooling layers. It is the job of the convolutional layers to collect high-level features from the input image, and the pooling layers' job to minimize the spatial dimension of those features. The output from the convolutional and pooling layers is flattened before being fed into a series of fully connected layers. The learned features are utilized as input to these fully linked layers, which then provide predictions. In the case of image classification, the final fully connected layer has 1000 units, each corresponding to a different class of objects. After each convolutional layer, the VGG19 model employs a rectified linear unit (ReLU), which serves as its activation function. This activation function allows the network to learn more complex and nonlinear features, which can improve its accuracy [16].

ResNet50

ResNet50 is a deep CNN that was introduced by researchers at Microsoft Research in 2015. It is a variation of the ResNet architecture and consists of 50 layers. Its purpose is to circumvent the issue of disappearing gradients that occurs in very deep neural networks. The following is a stepwise explanation of how ResNet50 functions:

1. The input to the model is a 224 × 224 RGB image.
2. The input image is passed through a series of convolutional layers with a stride of 2 to reduce the spatial dimension of the input.

3. The output of the convolutional layers is passed through a series of residual blocks, which are designed to help the network learn to make better use of the information from earlier layers.
4. Each residual block consists of two convolutional layers with batch normalization and a shortcut connection that bypasses the convolutional layers.
5. The shortcut connection is used to add the output of the convolutional layers to the input of the block, so enabling the network to acquire knowledge about the difference between the input and the block's output as a residual.
6. The output of each residual block is passed through a series of convolutional layers and average pooling layers to reduce the spatial dimension of the features.
7. The output is then flattened and passed through a series of fully connected layers to make predictions.
8. The final fully connected layer has 1000 units, each corresponding to a different class of objects.
9. Stochastic gradient descent combined with a cross-entropy loss function is used throughout the training process of the network.

The use of residual blocks helps to make the network easier to train and more accurate, even when it has many layers [17].

InceptionV3

Inceptionv3 is a CNN architecture that was introduced by researchers at Google in 2015. It is designed to be both deep and computationally efficient, by using a combination of parallel and hierarchical convolutions. The following is a stepwise explanation of how Inceptionv3 functions:

1. The input to the model is a 299 × 299 RGB image. In the research it is converted into 224 × 224.
2. The input image is passed through a series of convolutional layers with different filter sizes, including 1 × 1, 3 × 3, and 5 × 5 filters.
3. In addition to the standard convolutional layers, Inceptionv3 also includes "Inception modules," which are designed to capture features at multiple scales.
4. Each Inception module consists of a set of parallel convolutions with different filter sizes, followed by a pooling layer.
5. To obtain the final output, the outputs of all of the Inception modules are concatenated and then processed via a series of convolutional layers.
6. Inceptionv3 also includes batch normalization and dropout layers to prevent overfitting and improve the network's generalization ability.
7. The final output is passed through a softmax function to produce class probabilities.

The Inceptionv3 use of parallel convolutions and Inception modules helps to make it both accurate and efficient, even when it has many layers [18].

DenseNet-121

DenseNet-121 is a CNN that was introduced by researchers at Facebook AI Research in 2017. It is designed to be both accurate and computationally efficient by using a dense connectivity pattern between layers. The following is a stepwise explanation of how DenseNet-121 functions:

1. The input to the model is a 224 × 224 RGB image.
2. The input image is passed through a set of convolutional layers with small 3 × 3 filters, followed by a batch normalization layer and a ReLU activation function.
3. Each convolutional layer is connected to all previous layers, creating a "dense block."
4. The output of each dense block is passed through a transition layer, which includes a 1 × 1 convolutional layer and a pooling layer to reduce the spatial dimension of the features.
5. A global average pooling layer receives the output of the last dense block to be processed, which averages the spatial dimensions of the features and produces a fixed-length feature vector.
6. To generate class probabilities, the output of the global average pooling layer is fed into a fully connected layer that is equipped with a softmax activation function.

The DenseNet-121 use of dense connectivity patterns helps to improve the flow of information through the network and reduce the number of parameters, making it both accurate and efficient [19].

The basic idea is to use the VGG19, ResNet50, Inceptionv3, and Densenet-121 models as feature extractors to extract relevant features from the medical images that may indicate the presence of NSCLC. The model is typically pre-trained on large-scale image datasets, such as ImageNet, and then fine-tuned on the medical image data to adapt it for the specific task of lung cancer detection. The fine-tuned VGG19 model can then be used to classify new medical images as either containing or not containing NSCLC. The output of the model can also be used to estimate the likelihood of NSCLC in the image.

When it comes to several computer vision tasks, including image classification and image segmentation, state-of-the-art results have been reached by the deep learning models VGG19, ResNet50, Inceptionv3, and Densenet-121.

PERFORMANCE METRICS

The effectiveness of every deep learning algorithm was assessed with regard to these four metrics: accuracy, sensitivity, specificity, and F1 score. Accuracy is a measurement of how well the algorithm performs overall in accurately identifying positive and negative instances. The ability of the algorithm to accurately recognize positive cases is what is measured by sensitivity, whereas the ability of the program to accurately recognize negative cases is what is measured by specificity [20]. Harmonic mean of recall and accuracy is determined by the F1 score, providing a more balanced measure of performance than accuracy alone [21]. The four deep learning models, VGG19, ResNet50, Inceptionv3, and Densenet-121, were trained and tested using a fivefold cross-validation approach.

RESULT AND DISCUSSION

The results presented in Figure 9.2 infer that Inceptionv3 performs best in NSCLC detection, with the highest accuracy, sensitivity, specificity, and F1 score. The results showed that all four models achieved high accuracy, with Inceptionv3 achieving the highest accuracy of 98.01%, followed by VGG19 (96.89%), ResNet50 (96.87%), and Densenet-121 (96.34%). Inceptionv3 achieved the highest F1 score of 0.954, followed by ResNet50 of 0.946, Densenet-121 (0.931), and VGG19 (0.913).

However, CT scans have limitations in detecting small nodules that may be missed by radiologists. False positives and false negatives can also occur, leading to unnecessary procedures or missed diagnoses.

The results of the comparative analysis indicate that deep learning algorithms such as VGG19, ResNet50, Inceptionv3, and Densenet-121 can be effective in NSCLC detection using CT scans. However, the performance of the algorithms varies, with Inceptionv3 performing best in this study. The superior performance of Inceptionv3 can be attributed to its use, which allows the network to learn more efficient feature representations of images. This is likely due to its use of multiple scales and filters. VGG19 and ResNet50 have a similar performance, likely due to their simpler architectures compared with Inceptionv3.

The approach of using Inceptionv3 for NSCLC detection is similar to using VGG19, ResNet50, and DenseNet-121. The model is pre-trained on large-scale image datasets, such as ImageNet, and then fine-tuned on the medical image data to adapt it for the specific task of lung cancer detection.

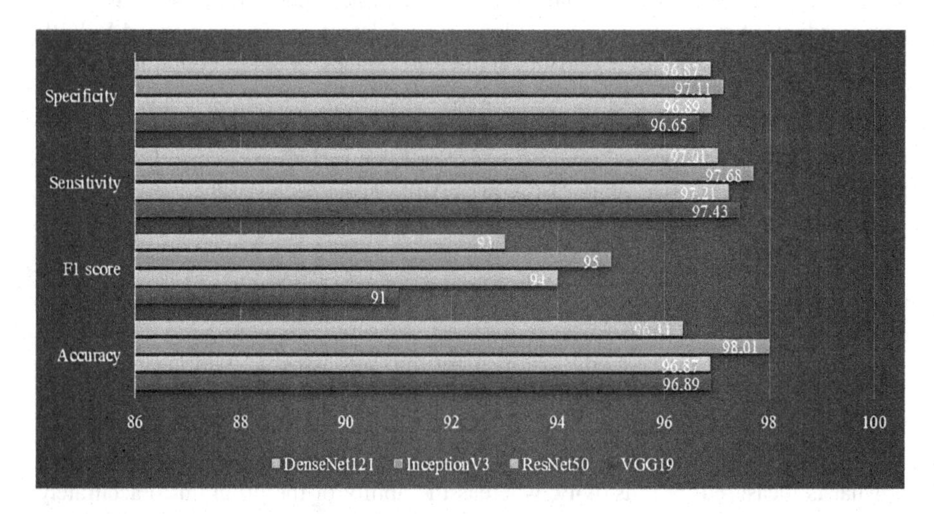

FIGURE 9.2 Comparative analysis of VGG19, ResNet50, Inceptionv3, and Densenet-121.

Although the use of deep learning models such as VGG19, ResNet50, Inceptionv3, and Densenet-121 holds great promise for improving the accuracy and efficiency of NSCLC detection in medical imaging, further research and validation are needed before these methods can be widely adopted in clinical settings. Additionally, it is important to note that medical imaging should always be interpreted by trained medical professionals, and deep learning models should be used as an aid to support clinical decision-making rather than a replacement for human expertise. As with any deep learning model, it is important to note that Inceptionv3 should always be used as an aid to support clinical decision-making rather than a replacement for human expertise.

CONCLUSION

NSCLC is a common and deadly disease that requires early detection for effective treatment. CT scans are commonly used for NSCLC detection but have limitations in detecting small nodules and can lead to false positives and false negatives. CNNs have shown promise in improving the accuracy of NSCLC detection and can analyze large datasets of CT scans more quickly and accurately than radiologists. The development of an automated system for the detection of NSCLC using a CNN can significantly reduce the workload of radiologists and improve patient outcomes. This study shows that deep learning models, specifically Inceptionv3, can be effective in NSCLC detection using CT scans. The remarkable accuracy, precision, recall, and F1 score values shown by the model point to the possibility of its use in primary care. To test the effectiveness of these models on bigger datasets and in clinical situations that are more representative of the real world, more study is required.

REFERENCES

1. WHO. "The Top 10 Causes of Death," Dec. 09, 2020. https://www.who.int/news-room/fact-sheets/detail/the-top-10-causes-of-death
2. S. B. Clark and S. Alsubait, "Non-Small Cell Lung Cancer – StatPearls – NCBI Bookshelf," Non-Small Cell Lung Cancer – StatPearls – NCBI Bookshelf, Sep. 05, 2022. https://www.ncbi.nlm.nih.gov/books/NBK562307/
3. American Lung Association, "Types of Lung Cancer," Lung Cancer Basics | American Lung Association, https://www.lung.org/lung-health-diseases/lung-disease-lookup/lung-cancer/basics/lung-cancer-types
4. S. P. Primakov et al., "Automated detection and segmentation of non-small cell lung cancer computed tomography images," Nature Communications, vol. 13, no. 1, Jun. 2022, doi: 10.1038/s41467-022-30841-3.
5. C. Kuo, C. Huang, J. Siao, C. Hsieh, V. Huy, K. Ko and H. Hsu, "Automatic lung nodule detection system using image processing techniques in computed tomography," Biomedical Signal Processing and Control, vol. 56, p. 101659, 2020, doi: 10.1016/j.bspc.2019.101659
6. G. M. M. Alshmrani, Q. Ni, R. Jiang, H. Pervaiz and N. M. Elshennawy, "A deep learning architecture for multi-class lung diseases classification using chest X-ray (CXR) images," Alexandria Engineering Journal, vol. 64, pp. 923–935, 2023, doi: 10.1016/j.aej.2022.10.053.
7. A. Narin, C. Kaya and Z. Pamuk, "Automatic detection of coronavirus disease (COVID-19) using X-ray images and deep convolutional neural networks," Pattern Analysis and Applications, vol. 24, no. 3, pp. 1207–1220, 2021, doi: 10.1007/s10044-021-00984-y.

8. Z. Xu, H. Ren, W. Zhou and Z. Liu, "ISANET: Non-small cell lung cancer classification and detection based on CNN and attention mechanism," Biomedical Signal Processing and Control, vol. 77, p. 103773, 2022, doi: 10.1016/j.bspc.2022.103773.
9. S. H. Kassania, P. H. Kassanib, M. J. Wesolowskic, K. A. Schneidera and R. Detersa, "Automatic detection of coronavirus disease (COVID-19) in X-ray and CT images: A machine learning based approach," Biocybernetics and Biomedical Engineering, vol. 41, no. 3, pp. 867–879, 2021, doi: 10.1016/j.bbe.2021.05.013.
10. D. Ardila et al., "End-to-end lung cancer screening with three-dimensional deep learning on low-dose chest computed tomography," Nature Medicine, vol. 25, no. 6, pp. 954–961, May 2019, doi: 10.1038/s41591-019-0447-x.
11. S. I. Priyadharsini, N. Mangayarkarasi, L. Sairamesh and G. Raghuraman, "Lung nodule detection on CT images using image processing techniques," International Journal of Pure and Applied Mathematics, vol. 119, no. 7, pp. 479–487, 2018.
12. N. Khehrah, M. S. Farid, S. Bilal and M. H. Khan, "Lung nodule detection in CT images using statistical and shape-based features," Journal of Imaging, vol. 6, no. 2, 2020, doi: 10.3390/jimaging6020006.
13. K. Sekeroglu and ÖM. Soysal, "Multi-perspective hierarchical deep-fusion learning framework for lung nodule classification," Sensors, vol. 22, no. 22, p. 8949, 2022, doi: 10.3390/s22228949.
14. X. Chen, Q. Duan, R. Wu and Z. Yang, "Segmentation of lung computed tomography images based on SegNet in the diagnosis of lung cancer," Journal of Radiation Research and Applied Sciences, vol. 14, no. 1, pp. 396–403, 2021, doi: 10.1080/16878507.2021.1981753.
15. The IQ-OTH/NCCD Lung Cancer Dataset. (n.d.). The IQ-OTH/NCCD Lung Cancer Dataset | Kaggle. https://www.kaggle.com/datasets/adityamahimkar/iqothnccd-lung-cancer-dataset
16. K. Simonyan and A. Zisserman, "Very deep convolutional networks for large-scale image recognition," arXiv.org, Apr. 10, 2015. https://arxiv.org/abs/1409.1556v6
17. K. He, X. Zhang, S. Ren and J. Sun, "Deep residual learning for image recognition," arXiv.org, Dec. 10, 2015. https://arxiv.org/abs/1512.03385v1
18. Z. Liu, C. Yang, J. Huang, S. Liu, Y. Zhuo and X. Lu, "Deep learning framework based on integration of S-Mask R-CNN and Inception-v3 for ultrasound image-aided diagnosis of prostate cancer," Future Generation Computer Systems, vol. 114, pp. 358–367, 2021, doi: 10.1016/j.future.2020.08.015.
19. G. Huang, Z. Liu, L. van der Maaten and K. Q. Weinberger, "Densely connected convolutional networks," arXiv.org, 2016. https://arxiv.org/abs/1608.06993.
20. S. Kumar, H. Kumar, R. Agarwal and V. K. Pathak, "Human disease prognosis and diagnosis using machine learning," Emerging Technologies for Computing, Communication and Smart Cities, Springer, Singapore, pp. 41–53, 2022, doi: 10.1007/978-981-19-0284-0_4.
21. S. Kumar and H. Kumar, "Lungcov: A diagnostic framework using machine learning and imaging modality," International Journal on Technical and Physical Problems of Engineering (IJTPE), vol. 14, no. 51, p. 2, June 2022. http://mail.iotpe.com/IJTPE/IJTPE-2022/IJTPE-Issue51-Vol14-No2-Jun2022/23-IJTPE-Issue51-Vol14-No2-Jun2022-pp190-199.pdf

10 Abnormality Classifications Using Machine Learning

Anupam Singh
School of Computer Science,
UPES Dehradun, UK, India

Ravendra Singh
Pranveer Singh Institute of Technology,
Kanpur, UP, India

Nitin Arora
Indian Institute of Technology,
Roorkee, Uttarakhand, India

INTRODUCTION

Information technology and e-healthcare systems have significantly improved the medical sector in the past few years. This chapter aims to address the challenge of separating central nervous system tissues from pathological dopaminergic neurons, such as gray matter (GM), white matter, and meninges (cerebrospinal fluid [CSF]), from magnetic resonance imaging (MRI) scans using a method for feature extraction and a support vector machine (SVM) classifier [1]. Brain tumors can be classified as benign or malignant, with fibroids being the most common type of benign tumor. Benign tumors are treatable and do not spread across the brain, while malignant tumors are cancerous and can invade different body regions. Early detection of tumors increases the patient's chances of survival, and MRI [2] is the most commonly used imaging technique for detecting brain tumors. It is noteworthy that MRI is not supported by the magnetic field principle; the data generated by an MRI structure is multidimensional, and the finding of brain cancers depends on segmentation, which differentiates the tumor zone from the background of an MR image. Brain tumors are defined as the unchecked growth of malignant cells in the brain, and brain tumors can be categorized into two groups: benign and malignant. Malignant brain tumors have an uneven and heterogeneous structure, and they contain active cancer cells, whereas benign brain tumors have a stable appearance and no aggressive cancer cells. Tumors that are categorized as low grade, including meningiomas and gliomas, are considered benign, whereas those that are classified as high grade, such as astrocytomas and glioblastomas, are deemed malignant.

DOI: 10.1201/9781003388319-10

Medical-based diagnosis places great value on MRI-based imaging technology. Multidimensional statistical observations make up MRI scans, and MRI provides targeted results that aid in treating illnesses and support pathologic cases to provide precise observations. By utilizing an optimum set of intelligent rules, segmentation [3] increases accuracy [4, 5]. Numerous practical sectors, like the manufacturing industry, health, signal processing, etc., have major uses for optimization [6, 7]. In straightforward computers or other gadgets with a constrained feature set, an optimized system can greatly improve memory use [8–10]. Software performance should always be reliable and independent of hardware constraints. Hence, the industry may benefit from improved solutions to provide clients with an approach that is simple and convenient. As a result, intelligence tools generated through simple optimization methods are important for advanced system intelligence.

The separation of the images can be accomplished in one way using thresholding. This method groups pixels into specific clusters according to luminous and perceptual qualities that are labeled according to the threshold level of picture data. Because it is difficult to precisely discern various heights over the image histogram, techniques based on the global histogram are typically recommended for setting the threshold rather than multimodal histograms. For labeling the blocks of parameterized and nonparametric image data, this thresholding is highly common. Luminous density patterns with Gaussian distribution patterns are present in the majority of parametric approaches with gray shade level analysis across each class. Such a method requires a significant amount of time and is computationally complicated. Nonparametric techniques determine the threshold level using a criterion associated with optimal results. When assets and time are limited, they provide compelling and complex dilemmas. Comparing the ecofriendly rule book to other existence improvement methods, the upside is the narrower tuning factor range that is needed. Due to connecting inside self-arranged construction, the technique associated with chaos theory is related to community decision. It is built on a population that evolved simple cooperative norms that were distributed regionally to one another and their unique environments. Solutions that are suitable develop as a result of swarm elements working together. Colonies, hen flocks, bacterial booms, cuckoo searches, schooling fish, and other biological systems are all included in the swarm intelligence approach [11–15]. Cuckoo's search algorithm as a result of some cuckoo species coercively parasitizing other host bird species by putting their eggs in their nests. Few host birds were able to engage the cuckoos in consistent fighting [16–18].

To detect brain cancers, a range of MRI sequences, are combined in the study. These different MRI sequences offer distinct information about the brain and tumor structure, helping practitioners locate the tumor accurately, determine its size, and understand how it affects nearby tissues. While benign brain tumors look more homogenous and don't have any active cancer cells, astrocytomas frequently have a non-uniform or diverse structure and do have active cancer cells. Making the distinction between squamous cell carcinoma is crucial for the diagnosis of brain tumors because it can influence the choice of a course of therapy. Prenatal recognition of brain tumors is essential for enhancing patient outcomes because it enables quick radiological evaluation and therapeutic decision-making. The optimal treatment plan, whether it be operation, cancer treatment, chemotherapy, or a mix of these approaches, can be determined by the location, size, and impact of the tumor.

We proposed an SVM-based prototype for rapidly and precisely finding anomalies in MRI data. The prototype was composed mostly of a feature extraction module and an SVM-based classification module. The feature extraction module extracted features from the MRI images, including statistical and textural features, using a range of image processing techniques. The SVM-based classification module then used these traits to split the MRI images into categories for normal and abnormal conditions.

BACKGROUND STUDY

Using clinical images like MRI and medical picture segmentation is an essential step in identifying and evaluating brain cancers (MRI). The procedure of image fragmentation involves breaking a picture into a number of regions or pieces, each of which matches to a distinct kind of structure or tissue in the image. The Expectation-Maximization (EM) algorithm, Expectation-Maximization (FCM), support vector algorithms, neural network algorithms, and other methods can all be used to segment medical images. These methods can be used to categorize various brain regions in MRI imaging as tumor- or non-tumor-bearing, which can assist medical practitioners in deciding on the best course of therapy for the patient. Accurately segmenting MRI images and identifying the size and location of brain tumors can aid healthcare providers in effectively planning and performing surgical procedures, chemotherapy, and radiation therapy, resulting in better patient outcomes. Our proposed method involves the application of two pixel-based segmentation techniques: k-means clustering and histogram analysis. The histogram analysis technique categorizes images pixel by pixel using one or more criteria. It locates the histogram's weakest point, typically located between two subsequent maxima, by searching for peak values in the histogram and using that information to establish the gray value blocking threshold [6, 7]. Histogram analytics is a method used to analyze data and can be useful in identifying bimodal histograms, whereas the k-means clustering method can be used to partition the data into categories based on similar or distinct feature vectors. The fusion-based approach involves overlaying a victim's train image on a test image to identify the tumor, but this method can be complex and time-consuming. Edge detection using the Canny method is a more efficient approach but does not support colored images and can take longer to identify the optimal solution.

Machine learning-based abnormality categorization has gained popularity in recent years due to its potential to enhance the diagnosis and management of a variety of medical disorders. Healthcare practitioners can utilize machine learning algorithms to analyze complex medical data and produce precise predictions of abnormalities, which can aid in their decision-making. Deep learning (DL), which makes use of neural networks with several hidden layers, is one of the most often used machine learning methods for classifying abnormalities. According to several studies [19–21], DL has been proven to be very successful at identifying and classifying a variety of medical diseases, including diabetes, heart disease, and cancer. With varied degrees of effectiveness, other machine learning methods such as SVMs, decision trees, and random forests have also been used to classify abnormalities [22–24].

The essential initiatives for classifying and segmenting brain tumors using MR images are briefly covered in this section. Clustering algorithms [10], unsupervised forests [11], choice of forests [12], analysis of principal components [13], and fuzzy random forests were some of the first machine learning techniques created as well as SVMs [10]. These algorithms frequently classify image voxels using hand-crafted features, which calls to select the most interesting traits from training photos, so a human specialist is needed. Recently, encouraging results have been obtained when segmenting medical pictures using DL-based techniques [14]. Because DL techniques can routinely learn complicated picture features from the training data, they enable us to create feature vectors that are more reliable. A variety of DL models have been developed for robotic tumor identification with positive outcomes [15]. Pereira and associates constructed two distinct two-dimensional (2D) CNNs with higher sections as a clustering classifier to isolate both low-grade and high-grade glioblastomas. The 2D CNN method, which integrates global and local information on large patches, offers two choices [16]. For instance, a local approach examines information in the immediate area, whereas a major approach examines the wider context of MRI. A densely integrated conditional random field (CRF) network was used in the comment step by [17] to analyze three-dimensional (3D) patches, two feeds, and globally relevant data. Tumor division can be improved by utilizing another more powerful version of DeepMedic that captures residual connections [4]. The aforementioned methods developed using CNNs for brain tumor classification take into account local areas in MRI images to categorize each patch of a brain tumor [25]. The center pixel is labeled in accordance with the classification results; as a result, it only looks into spatially constrained contextual data.

Machine learning has a wide range of applications, including the classification of abnormalities in sensor data, electroencephalography (EEG), and medical imaging. Research in this field has focused particularly on medical imaging because it is possible to accurately diagnose illnesses like cancer and neurological disorders by analyzing complex medical images using machine learning [26, 27]. Using machine learning to classify abnormalities has a number of obstacles, one of which is the requirement for big, high-quality datasets. Obtaining such datasets can frequently be challenging because they may call for substantial data collection and annotation. Furthermore, it might be tricky to comprehend how machine learning models produce their predictions because they are frequently sophisticated and challenging to interpret. Despite these difficulties, machine learning-based anomaly classification has demonstrated considerable potential for enhancing medical diagnosis and therapy. It is expected that machine learning techniques will play a bigger part in medical research and practice as they advance and become more sophisticated.

METHODOLOGY

The following is a general outline of the methodology section:

- Data collection: Describe the dataset that was used in the investigation, mentioning its origin, size, and data type. Describe any data cleaning procedures used to ensure the data's quality as well as how the dataset was preprocessed.

- Feature extraction: Describe the feature extraction process, including the features that are extracted and the feature selection techniques. Describe the features' transformation and normalization processes so they can be used by the machine learning algorithm.
- Machine learning algorithm: Include the type of method (supervised, unsupervised, etc.) and the exact algorithm employed (logistic regression, decision trees, neural networks) when describing the machine learning algorithm used to classify abnormalities. Describe the training, testing, and validation of the method, including any hyperparameter adjustment.
- Model evaluation: Describe the metrics, such as accuracy, precision, recall, F1 score, and area under the receiver operating characteristic (ROC) curve that are used to assess the performance of the machine learning model. Describe the model's evaluation process on the training and test datasets and provide the evaluation's findings.
- Comparison with other methods: Compare the effectiveness of the created machine learning model against alternative anomaly classification techniques, including established statistical techniques or other machine learning algorithms. Discuss the generated model's advantages and disadvantages while making suggestions for enhancements.
- Ethical considerations: Describe any ethical factors, such as patient privacy and data confidentiality that were taken into account during the development and testing of the machine learning model.

Brain tumors have been detected and classified using the suggested method using an image processing technology. The suggested approach entails a number of stages to achieve the desired outcomes. The first phase involved acquiring an MRI image, and the following step involved preprocessing the image. In the subsequent stage, the picture is separated using k-mean clumping. In the final step, SVM [8–10] is applied to predict accuracy. The next sections discuss each of these actions.

Phase1: Pre-processing

An MRI picture is initially taken to prepare it to serve as an input for the algorithm. This stage is followed by the 2D image's scaling [11, 12]. A picture is reshaped to become homogeneous by being changed into a dimension with the same size. The technique then moves on to grayscale photo conversion from a double format [13, 14]. This conversion's goal is to simplify a common RGB image's complexity. The flow graph of the proposed system for brain tumor detection is presented in Figure 10.1.

Phase 2: Feature Retrieval

To locate the terminal disease, the brain MR pictures must be split into two parts [15, 16]. The tumor nerve cells in the brain are focused in one region, while the healthy brain cells are concentrated in another [17]. The MR image, which holds a lot of important information, was subjected to the feature separation process, and key components were removed from an image that precisely described the MR image.

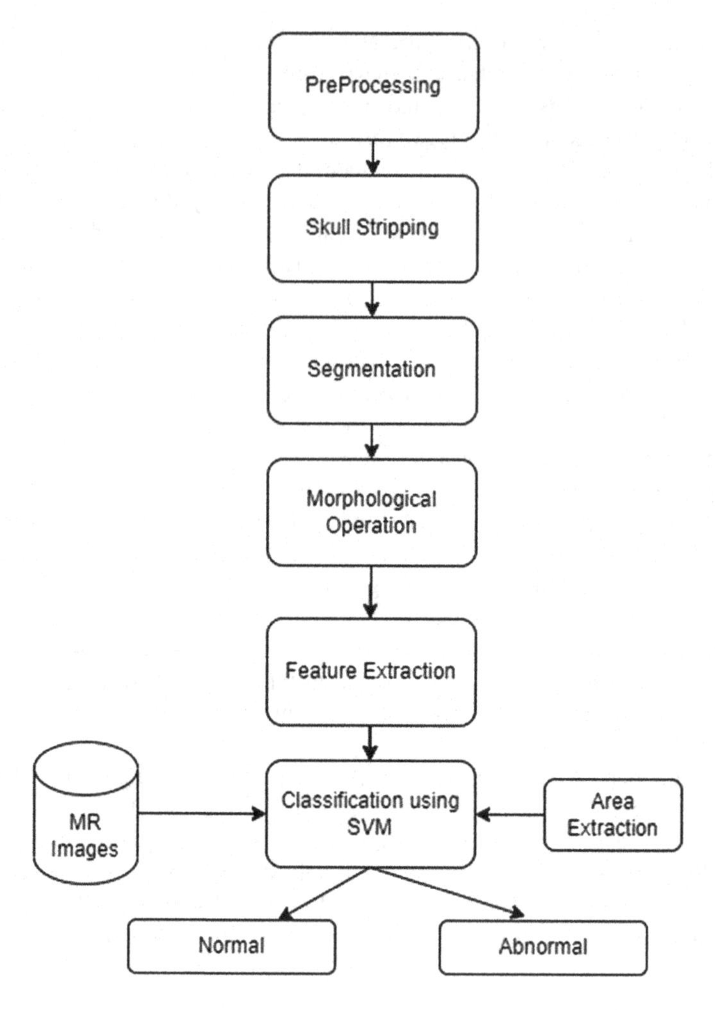

FIGURE 10.1 Proposed model of classification.

The suggested approach to extracting features from an image uses the discrete wavelet transform (DWT) algorithm. Both time and amplitude information are revealed by DWT. As MR images are two-dimensional, 2D-DWT [25, 28, 29] is used for MR imaging. The high and low passes of the filters are used by the DWT. Both filters were applied in the first stage, along with the columns of the downscaled MR picture. The bandwidth of the MR picture is cut in half after downsampling, leaving only the necessary characteristics. The Nyquist criterion reinforces the input signals from incomplete information, maintaining the details of the MR image. Following downsampling and MRI row rows, the low-pass and high-pass screens are used. While the lowest-pass filters in the DWT [3, 5, 18, 25, 30] produce an estimate of MRI, high-pass scrubbers in the DWT include an edge of the MR picture. The first band acquired, the LL sub-band, closely resembles MR pictures. The HL sub-band is used to restore an image's vertical components, and the LH thread is used to retrieve an MR image's horizontal characteristics. The diagonal properties of the MR picture are derived from HH.

Phase 3: Classification

To classify an input test image, it is necessary to divide it into two distinct categories, normal and abnormal, following an evaluation of its generalization capability with k-nearest neighbors (KNN). SVM is employed for categorization, and the image displays the typical and normal versions of the secret image. SVM is a recent and highly effective learning tool in supervised categorization technology, surpassing RBF algorithms and neural networks in efficiency. Arora et al. [4] used SVM for diabetes patients' classification. SVM employs an ideal linear separation hyperplane to separate two different types of data in subspace, and the smallest margin between two sets is maximized to obtain the perfect hyperplane. Therefore, the resulting hyperplane will only use boundary models that have undergone training, such as support vectors. SVM operates by nonlinearly transferring a training set to a tall space concealed from both I/P and O/P. Figure 10.2 illustrates the creation of the optimal hyperplane for distinguishing features. Finally, an SVM's conclusion is a square matrix of training data that is thrown onto a large feature space using a kernel function [25].

The structured risk reduction principle of statistical learning theory serves as the foundation for SVM. By managing empirical risk and classification capability, its main objective is to narrow the gap between classifications and the actual cost [31–33].

This method uses SVM with a linear parameter and RBF to divide images into "Normal" and "Abnormal" categories. These classes are designated with the letters "2" and "1," which stand for normal and abnormal, respectively. Medical diagnosis decisions about pattern recognition or classification have effects beyond the statistical metrics of accuracy and validity. The clinical or pathological interpretation of

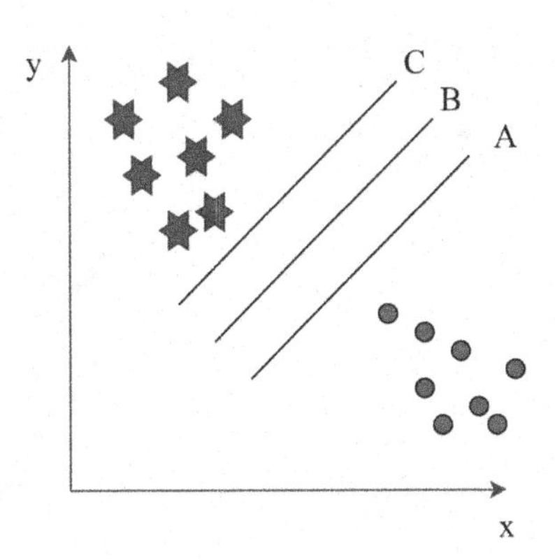

FIGURE 10.2 Hyperplanes (A, B, and C).

template vector-based empirical or rule-based judgment is required. These are the alternatives: a true positive (TP), sometimes known as a "hit," which indicates that the patient actually has the illness, and true negatives (TNs), which are tests that come back negative in a subject who does not have the ailment. A test is said to be falsely negative (FN) or to have "missed" the case when it comes to a patient who has the ailment. A misdiagnosis false alarm is when a test produces a positive result even though the subject is null for the syndrome.

RESULT DISCUSSION

In this study, the researchers suggested an SVM-based prototype for quickly and accurately locating anomalies in MRI data. A feature extraction module and an SVM-based classification module made up the prototype's two primary parts. Using a variety of image processing approaches, the feature extraction module extracted features from the MRI images, including statistical and texture features. These characteristics were then utilized by the SVM-based classification module to divide the MRI pictures into normal and pathological categories.

To categorize the test input image, the KSM classifier is utilized. Figure 10.2 illustrates the detection of region of interest (ROI) in an image belonging to the anomalous class. Figure 10.3 depicts the hyperplane in 2D and 3D. To determine the categorization accuracy, a confusion matrix was employed. Figure 10.4 illustrates the difference between original abnormality and GM removal. Figure 10.5 shows the SVM (linear) and SVM (RBF) confusion matrix.

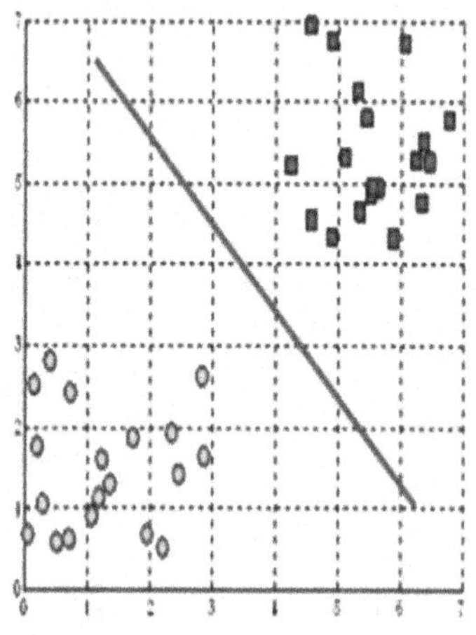

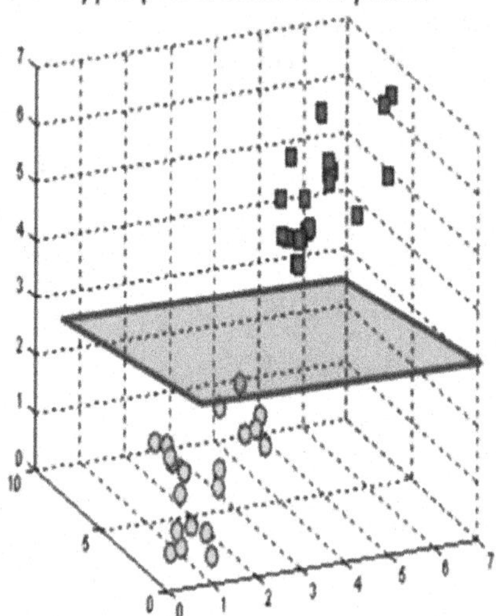

FIGURE 10.3 Hyperplanes in 3D and 2D feature space.

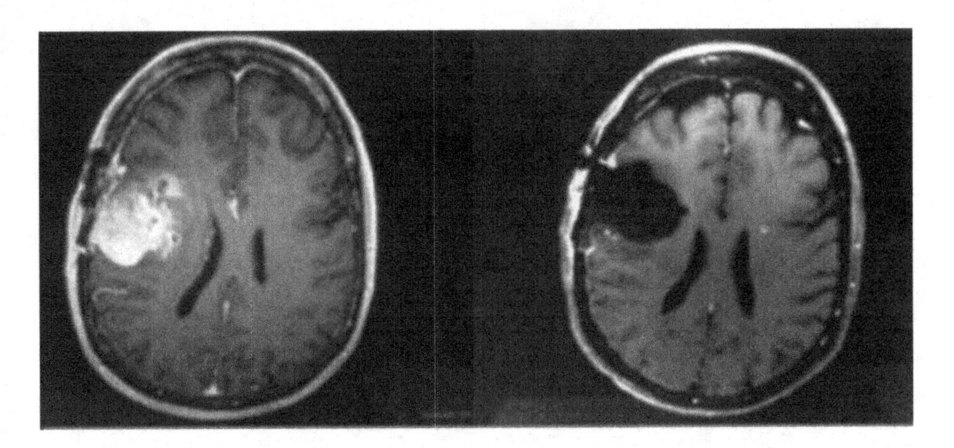

FIGURE 10.4 Abnormality detection.

	Abnormal	Normal	Total
Normal	6	22	28
Abnormal	16	11	27

(a)

	Abnormal	Normal	Total
Normal	3	23	26
Abnormal	22	3	25

(b)

FIGURE 10.5 Confusion matrix of (a) linear kernel SVM and (b) RBF kernel SVM.

CONCLUSION

SVMs are frequently employed in MRI and other medical imaging applications for classification and anomaly identification. By identifying the optimum hyperplane that divides the data points into several classes, SVMs, a common machine learning technique, divide the data points into various classes. SVMs can handle high-dimensional feature spaces with relatively small training sets, making them particularly effective in scenarios where there are many features and few data. This study offers an SVM-the based prototype for object identification that can identify anomalies in MRI images in true time and with high accuracy. The study suggests a method for locating abnormalities in MRI pictures using an SVM classifier, which has a 90% accuracy rate for classifying incoming image data as normal or abnormal. The suggested system has the potential for expansion to detect anomalies in multimodal brain images and enhance the classification accuracy. Furthermore, the system can be customized for different applications, irrespective of image size, support vector count, or window size. By developing accurate and efficient object recognition systems using machine learning algorithms like SVMs, healthcare professionals can improve their ability to detect and diagnose brain tumors and other abnormalities in medical imaging modalities. This can ultimately lead to better treatment outcomes for patients. The SVM-based prototype used in this study successfully detected anomalies with a high degree of accuracy, illustrating the potential of machine learning methods like SVMs for anomaly identification in medical imaging applications. Real-time MRI anomaly detection can aid in the early detection and treatment of a number of illnesses, including brain tumors, strokes, and neurodegenerative diseases. To assess the SVM-based prototype's performance on bigger datasets and explore its potential for usage in clinical settings, more research is required in this area.

REFERENCES

1. Nasr-Esfahani, M., Mohrekesh, M., Akbari, M., Soroushmehr, S. R., Nasr-Esfahani, E., Karimi, N., Samavi, S., & Najarian, K. (2018, July). Left ventricle segmentation in cardiac MR images using fully convolutional network. In 2018 40th Annual International Conference of the IEEE Engineering in Medicine and Biology Society (EMBC), pp. 1275–1278.
2. Kumar, S., Dabas, C., & Godara, S. (2017). Classification of brain MRI tumor images: A hybrid approach. Procedia Computer Science, 122, 510–517.
3. Hiremath, P., Humnabad Iranna, Y., & Pujari, J. D. (2007). Classification of squamous cell carcinoma based on color and textural features in microscopic images of esophagus tissues. Journal of Computer Science, 3(7), 566–573.
4. Arora, N., Singh, A., Al-Dabagh, M. Z. N., & Maitra, S. K. (2022). A novel architecture for diabetes patients' prediction using K-means clustering and SVM. Mathematical Problems in Engineering, 2022, 4815521.
5. Yeh, J. Y., & Fu, J. C. (2008). A hierarchical genetic algorithm for segmentation of multi-spectral human-brain MRI. Expert Systems with Applications, 34(2), 1285–1295.
6. Godara, S., Singh, R., & Kumar, S. (2016). A novel weighted class based clustering for medical diagnostic interface. Indian Journal of Science and Technology, 9(44), 1–11.
7. Iqbal, A., & Jeoti, V. (2012). A novel wavelet-Galerkin method for modeling radio wave propagation in tropospheric ducts. Progress in Electromagnetics Research B, 36, 35–52.
8. Singh, K., & Verma, S. (2012). Detecting brain MRI anomalies by using SVM classification. International Journal of Engineering Research and Applications (IJERA), 2, 724–726.
9. Singh, R., & Agarwal, B. B. (2023). An automated brain tumor classification in MR images using an enhanced convolutional neural network. International Journal of Information Technology, 15, 665–674.
10. Singh, R., & Agarwal, B. B. (2022). A hybrid approach for detection of brain tumor with levy flight cuckoo search. Webology, 19(1), 5388–5401.
11. Singh, R., & Agarwal, B. B. (2022). Abnormality detection and classification from brain MRI using machine learning. International Journal of Health Sciences, 6(S3), 9170–9180. https://doi.org/10.53730/ijhs.v6nS3.8242.
12. Najadat, H., Jaffal, Y., Darwish, O., & Yasser, N. (2011). A classifier to detect abnormality in CT brain images. In Proceeding of the International Multiconference of Engineers and Computer Scientists, pp. 16–18.

13. Hiremath, P. S., Bannigidad, P., & Hiremath, M. (2011). Digital microscopic image analysis of virus particles. International Journal of Machine Intelligence, 4, 180–184.
14. Qurat-Ul-Ain, Latif, G., Kazmi, S. B., Jaffar, M. A., & Mirza, A. M. (2010). Classification and segmentation of brain tumor using texture analysis. Recent advances in artificial intelligence, knowledge engineering and data bases. In Proceedings of the 9th WSEAS International Conference on Recent Advances in Artificial Intelligence, Knowledge Engineering and Data Bases, pp. 147–155.
15. Kharrat, A., Gasmi, K., Ben Messaoud, M., Benamrane, N., & Abid, M. (2010). A hybrid approach for automatic classification of brain MRI using genetic algorithm and support vector machine. Leonardo Journal of Sciences, 17(1), 71–82.
16. Ranjan, R., Singh, A., Rizvi, A., & Srivastava, T. (2020). Classification of Chest Diseases Using Convolutional Neural Network. In Singh, P., Pawłowski, W., Tanwar, S., Kumar, N., Rodrigues, J., Obaidat, M. (eds). Proceedings of First International Conference on Computing, Communications, and Cyber-Security (IC4S 2019). Lecture Notes in Networks and Systems, vol 121. Springer, Singapore.
17. Rejani, Y., & Thamarai Selvi, S. (2009). Early detection of breast cancer using SVM classifier technique. arXiv preprint, arXiv:0912-2314.
18. Gao, Z., Po, L., Jiang, W., Zhao, X., & Dong, H. (2007). A novel computerized method based on support vector machine for tongue diagnosis. In 2007 Third International IEEE Conference on Signal-Image Technologies and Internet-Based System, pp. 849–854. IEEE.
19. Gao, M., Bagheri, M., & Dengel, A. (2018). Deep learning for automatic glaucoma diagnosis based on multiple ocular features. Journal of Glaucoma, 27(8), 775–781.
20. Rajkomar, A., Dean, J., & Kohane, I. (2018). Machine learning in medicine. New England Journal of Medicine, 379(14), 1378–1379.
21. Wang, S., Kang, B., Ma, J., Zeng, X., Xiao, L., Guo, J., & Tian, J. (2019). A deep learning algorithm using CT images to screen for Corona virus disease (COVID-19). European Radiology, 31(8), 6096–6104.
22. Abbas, Q., Aziz, W., Munir, A., & Mahmood, T. (2014). A comparative study of different machine learning techniques for abnormality detection in medical images. Procedia Computer Science, 37, 391–398.
23. Alomari, R. S., Bawadi, M. M., & Al-Handal, A. H. (2015). Classification of medical images using machine learning techniques. Journal of Computer Science, 11(9), 910–919.
24. Fathi, A., Abdolahi, M., Rahimpour, A., & Chavoshi, M. (2016). Comparison of various machine learning techniques in mammography abnormality classification. Journal of Medical Signals and Sensors, 6(2), 102–111.
25. Zhang, Y.-D., & Wu, L. (2008). Weights optimization of neural network via improved BCO approach. Progress in Electromagnetics Research, 83, 185–198.
26. Chen, P. H., Liu, C. L., Yang, B. H., Huang, Y. W., & Huang, C. K. (2019). Deep learning for brain tumor classification: A comprehensive review. Computers in Biology and Medicine, 109, 134–153.
27. Esteva, A., Kuprel, B., Novoa, R. A., Ko, J., Swetter, S. M., Blau, H. M., & Thrun, S. (2019). Dermatologist-level classification of skin cancer with deep neural networks. Nature, 542(7639), 115–118.
28. Xu, Y., Guo, Y., Xia, L., & Wu, Y. (2008). A support vector regression-based nonlinear modeling method for SiC MESFET. Progress in Electromagnetics Research, 2, 103–114.
29. Kumar, R., Singh, A., Datta, G., Kumar, A., & Garg, H. (2021). Brain tumor detection system using improved convolutional neural network, 2021 Sixth International Conference on Image Information Processing (ICIIP), Shimla, India, pp. 126–130, doi: 10.1109/ICIIP53038.2021.9702648.

30. Selvaraj, H., Thamarai Selvi, S., Selvathi, D., & Gewali, L. (2007). Brain MRI slices classification using least squares support vector machine. International Journal of Intelligent Computing in Medical Sciences & Image Processing, 1(1), 21–33.
31. Mancas, M., Gosselin, B., & Macq, B. (2005). Tumor detection using airways asymmetry. Engineering in Medicine and Biology Society. IEEE- EMBS 2005, 27th Annual International Conference, pp. 6528–6531.
32. George, E. B., Rosline, G. J., & Rajesh, D. G. (2015). Brain tumor segmentation using cuckoo search optimization for magnetic resonance images. In 2015 IEEE 8th GCC Conference & Exhibition, pp. 1–6.
33. George, E. B., & Karnan, M. (2012). MR brain image segmentation using bacteria foraging optimization algorithm. International Journal of Engineering and Technology, 4(5), 295–301.

11 Multilayer Perceptron-Based Speech Emotion Recognition for Identifying the Problematic Skills of Dyslexic Learners

Sarthika Dutt, Rohit Kanauzia, and Aditi
COER University, Roorkee, Uttarakhand, India

Himanshu Bartwal
Parnveer Institute of Technology, Kanpur, UP, India

INTRODUCTION

A learner with learning disability has different requirements to learn efficiently. To stay focused and involved in learning without losing attention, a learner with a learning disability has different requirements (Jacob et al., 2021). It is all too often forgotten that there are large numbers of people with cognitive disabilities. We recognize that given the huge range of abilities that it is nearly impossible to cater to all learning impairments. However, there are some general techniques that can help open up a site to as wide an audience as possible without becoming overly burdensome on the design and development of an application. These general techniques and applications are provided to the learners according to their learning and problematic skills (Dutt et al., 2021). Several disabilities have been analyzed and assistive tools and supports have been provided for the same. Assistance and support for the learning disabled and their inclusive education has not been the focus of previous research (Kapoor et al., 2018).

Dyslexia is a learning disorder that may be genetic and sometimes may occur naturally in people of any age. It is basically a language disorder, which means that the mind of people who suffer from dyslexia have difficulty in decoding language in any form, whether verbal or written (Habib, 2021). Dyslexia directly affects a person's ability to read, spell, write, and speak. The reading ability, fluency, phonological awareness, speech, and random naming are problematic skills for dyslexic learners (Snowling & Melby-Lervåg, 2016). Therefore, speech recognition can be effectively used to identify different learning skills of these learners. Also, determining the

DOI: 10.1201/9781003388319-11

emotions of learning-disabled learners through speech can provide useful insights to help them (Matre & Cameron 2022).

Therefore, in the present study speech emotions of learning-disabled learners have been analyzed to identify their problematic skills. Also, emotions play an extremely important role in human mental life. It is a medium of expression of one's perspective or one's mental state to others. Speech emotion recognition (SER) is the act of attempting to recognize human emotion and affective states from speech (Abbaschian et al., 2021). This capitalizes on the fact that voice often reflects underlying emotion through tone and pitch. There are few universal emotions, such as calm, anger, happiness, and sadness, in which any intelligent system with finite computational resources can be trained to identify or synthesize as required. The approach for SER primarily comprises two phases known as feature extraction and features classification phase (Byun & Lee, 2021). In the field of speech processing, researchers have derived several features such as source-based excitation features, prosodic features, vocal traction factors, and other hybrid features. The second phase includes feature classification using linear and nonlinear classifiers. In this work a model using MLPClassifier to recognize emotion from speech is utilized to extract features and a sound file library to read the recorded sound file of dyslexic learners. Spectral and prosodic features are used for SER because both of these features contain the emotional information of the learning-disabled learners (Song & Wei, 2021). The potential features are extracted from each utterance for the computational mapping between emotions and speech patterns. Pitch can be detected from the selected features, Radial basis function and backpropagation network are used to recognize the emotions based on the selected features, and it has been proved that radial basis function produces more accurate results for emotion recognition than the backpropagation network (Daneshfar et al., 2020). Emotion analysis has become a critical topic in various research communities, including natural language processing (NLP), Web mining, information retrieval, and human-computer interaction. The objective of this study is to predict the emotion of learning-disabled learners through audio/speech data and identify their strengths and weakness. This chapter is organized as follows: Literature Review, followed by Methodology, Results and Discussion, and finally the conclusion.

LITERATURE REVIEW

AI-Based Assistive Technologies for Solving Disabilities

Students with learning disabilities often have difficultly reading more advanced texts. They may not be able to follow complex sentence structures, or they could struggle with popular idioms found in the text itself. Scientists and researchers are molding artificial intelligence (AI) that can make these harder texts into more understandable resources (Dąbrowska, 2015). AI might create simpler sentences or replace popular quips with plainer alternatives. It should be a great way for students with learning disabilities to better relate to and engage with the material. Even educators can have more data. With AI, educators can have access to a variety of data, which can assist them in identifying the weakness among students. The promise of a more personalized education with the use of AI is appealing (Eginli & Özmelek Taş,

2018). Additionally, academics may become much simpler for students with learning disabilities who have historically struggled to grasp the material and keep up with their peers in an academic setting.

AI voice-assisted technologies, like Echo, Google Home, and Alexa, have created new means of accessibility for disabled people. As AI takes an important role in communication and interaction, the use of this technology enables individuals with disabilities to access information much easier, all just by speaking to their devices. AI technology can empower people living with limited physical mobility. Microsoft's AI for Accessibility program uses the potential of AI to develop solutions to many physical and cognitive challenges disabled individuals face at work and in daily life to promote social inclusion for them. Microsoft's initiative aims to increase independence and productivity for disabled people in employment, daily life, and communication.

Speech-Based Assistance for Disabled People

Dyslexic learners could benefit from speech-based assistive technologies to improve their communication and manage their work environment. The Emotional Sensitivity Assistance System (EmotAsS) has been introduced in a study for people with (mental, neurological, and cognitive) disabilities (Hantke et al., 2017). Advances in speech-to-text and text-to-speech technologies help aid those living with speech impediments. Voice systems like Voiceitt can learn speakers' pronunciation over time and translate the user's words into clear and normalized speech in the form of audio or text messages. The app may also be used to help people with speech impairments communicate face-to-face with each other (El Hammoumi et al., 2018). Similarly, Google's Parrotron is another AI tool that makes it easier for speech-impaired users to be understood by turning impaired speech patterns into fluent speech (Cave & Bloch, 2021). Emotions of dyslexic learners have not been analyzed to classify their problematic skills. In this study the emotions through speech data of learning-disabled learners are analyzed to provide required assistance.

It has been investigated in studies that learning disabilities have been identified utilizing machine learning (ML) or deep learning technologies, but speech recognition has not been considered for identifying dyslexia in children. Therefore, in the present study speech of dyslexic learners has been analyzed to identify their problematic skills while reading, and focusing on their speech data and emotions could provide useful insights to improve their learning. The procedure of the proposed study is discussed in detail in the next section.

METHODOLOGY

This section discusses the data collection, features section, feature extraction, and classification process involved in the proposed study.

Dataset Description

There was a total sample of 19 learners, 9 dyslexic learners and 10 non-dyslexic learners. These learners were approached either at Doon Psychotherapeutic Center or were at their registered government and private schools in the Dehradun District of the Uttarakhand State of India. The average age of the children who participated in this study was 11 years. In-person interviews were conducted with pretest questionnaires. Audio speech data was also recorded for the same and comprises 105 speech audio files with the. wav file extension. The same audio files were evaluated by 247 people about 10 times to verify the labels allotted to the audio files regarding emotions of the learning-disabled learners. Labels from 1 to 8 are used for the classification process; odd numbers indicate male speech data and even numbers indicate female speech data in the dataset.

Feature Extraction

A pretest has been conducted of 19 learning and non-learning-disabled learners with 19 pretest questionnaires, each mapped with a problematic skill of the dyslexic learner. Also, the audio/speech data of these learners have been recorded during the pretest to analyze their reading skills and emotions expressed by the learners during the pretest through recorded data.

The proposed methodology is depicted in Figure 11.1. Initially the dataset (including audio speech data) of learning- and non-learning-disabled learners was collected through surveys and in-person interviews. The features have been extracted utilizing mel frequency cepstral coefficients (MFCC), the mel scale, and chroma techniques, and finally classified utilizing MLPClassifier to generate profile details of the dyslexic learners.

- MEL Frequency Cepstral Coefficients (MFCC)

  MFCC is a feature in audio data that is extracted and classified in several studies. It is generated from the audio signals by dividing the signal into overlapping frames to apply fast Fourier transformation and a filter bank is applied to remove noise and cepstral coefficients.
- Chroma

  The chroma feature is used to collect features from audio signals or speech data; it is generally based on pitch classes. The pitch classes are classified into 12 categories, consisting of two factors: chroma vector and chroma-deviation

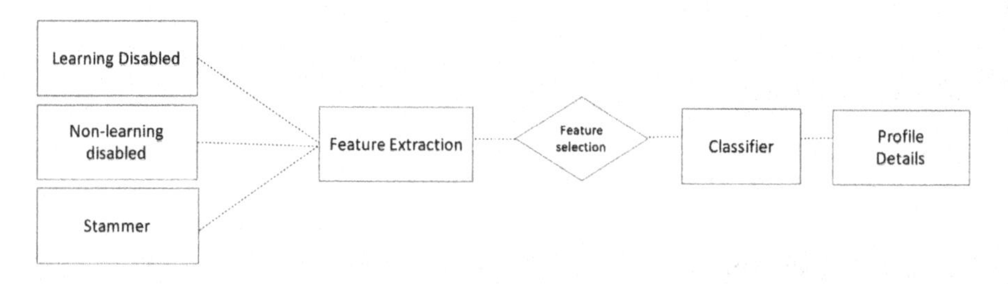

FIGURE 11.1 Proposed methodology workflow.

- Mel
  The mel scale is used to assess real recurrence of an unadulterated tone and relates its evident repeat, or pitch. People are efficient at perceiving little changes in pitch at low frequencies rather than high frequencies.
- Multilayer Perceptron (MLP) Classifier
  Multilayer perceptron (MLP) classifier is a class of feedforward artificial neural network (ANN). It is loosely for feedforward ANN and strictly for network of multiple layers of perceptron. The inputs (features) are served by the input layer and cross several hidden layers for final classification result at the output layer.

Procedure

The workflow is shown in Figure 11.2.

In the present study, MLPClassifier has been utilized for the classification process. It uses a supervised learning technique called backpropagation for training. MLP can be used to extract patterns and detect trends that are too complex for manual classification. This can be used to distinguish data that are not linearly separable and to separate the filtered data using more complex shapes (Sarker, 2021). This model can be applied to data with complex structures and it is comprised of an internal neural network for classification process.

SER is an act of attempting to recognize human speech and extract included emotions with affective states from the speech. This means that voices of people often reflect their emotions through their pitch and tone. This work uses a model using MLPClassifier to recognize emotion from speech using the librosa library to extract features and a sound file library to read the sound file. The emotional information in

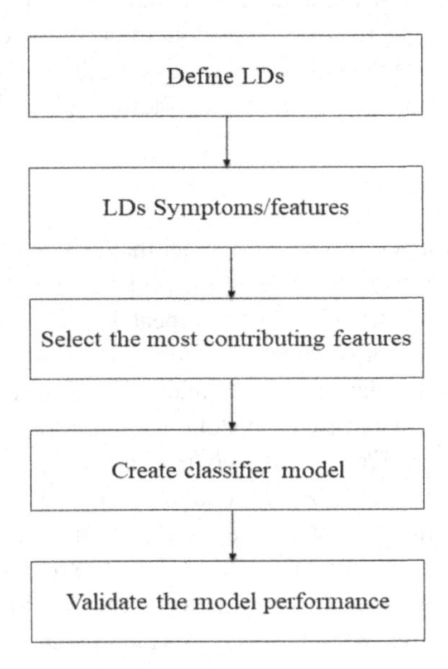

FIGURE 11.2 Workflow of SER using MLP classifier.

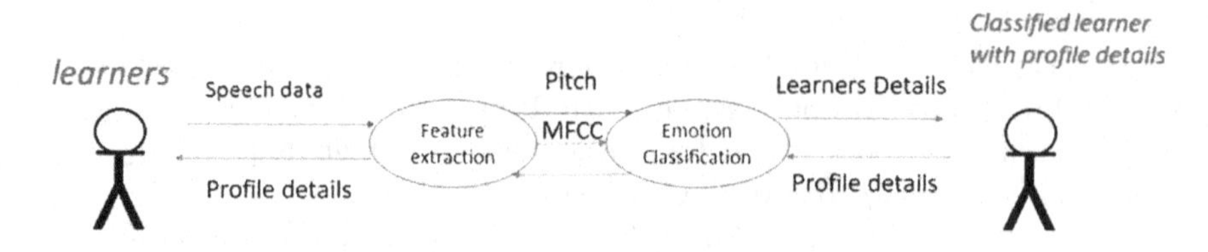

FIGURE 11.3 Speech emotion recognition system.

the voice data is contained by the spectral and prosodic features. In the speech recognition process (Figure 11.3), the input is audio speech data, which is preprocessed to transform it into acceptable format. The steps included in the speech data preprocessing are framing, hamming, windowing, etc. These steps are utilized to process audio files and to convert them into numerical values for training purposes. These numerical values represent the voice parameters or features, such as frequency, time, voice amplitude, and several other parameters. Finally, after extracting the features from audios, the focus is to train the extracted data using MLPClassifier. Data of the learners are comprised of the speech data of 19 learning and non-learning-disabled learners. The transformed numerical values or features of the learner's voice are then trained and stored with the numerical value of emotions in different arrays. These arrays were the inputs to the MLP classifiers used to detect different classes into which the speech of the learners falls under to generate learner profile.

The speech emotion detection system is implemented as an ML model. The flowchart (discussed previously) represents a pictorial overview of the process. The first step is data collection, which is of prime importance. The model being developed learned from the data given to it and all the decisions and results that a developed model produce are guided by the data. The second step, called feature engineering/selecting most contributing features, is a collection of several ML tasks that are executed over the collected data. These procedures address several data representation and data quality issues. The third step is often considered the core of an ML.

RESULT AND DISCUSSION

This model uses an ML algorithm to learn about the data and train itself to respond to any new data to which it is exposed. Finally, the proposed model is evaluated using several evaluation metrics. Very often, developers repeat the steps of developing a model and evaluating it to compare the performance of different algorithms. A total number of files for the three emotions in the dataset are 192 audio files for Excited, 192 audio files for Sad, and 192 audio files for Confused. The total number of features and emotions to be extracted from the dataset is 576. The total count of all emotions is given in Figure 11.4.

The librosa library will be utilized to load, display, and preprocess the audio files to extract the MFCC features. For now, we use it just for loading into a NumPy array. The next step would be to use the MFCC function to extract those features.

In SER, the role of feature extraction is to extract audio parameters to analyze learners' emotions. Features extraction is performed using MFCC, Chroma, and the mel spectrogram in the proposed study. In Python we have utilized librosa built-in functions for feature extraction and stored values in arrays for classifying learners.

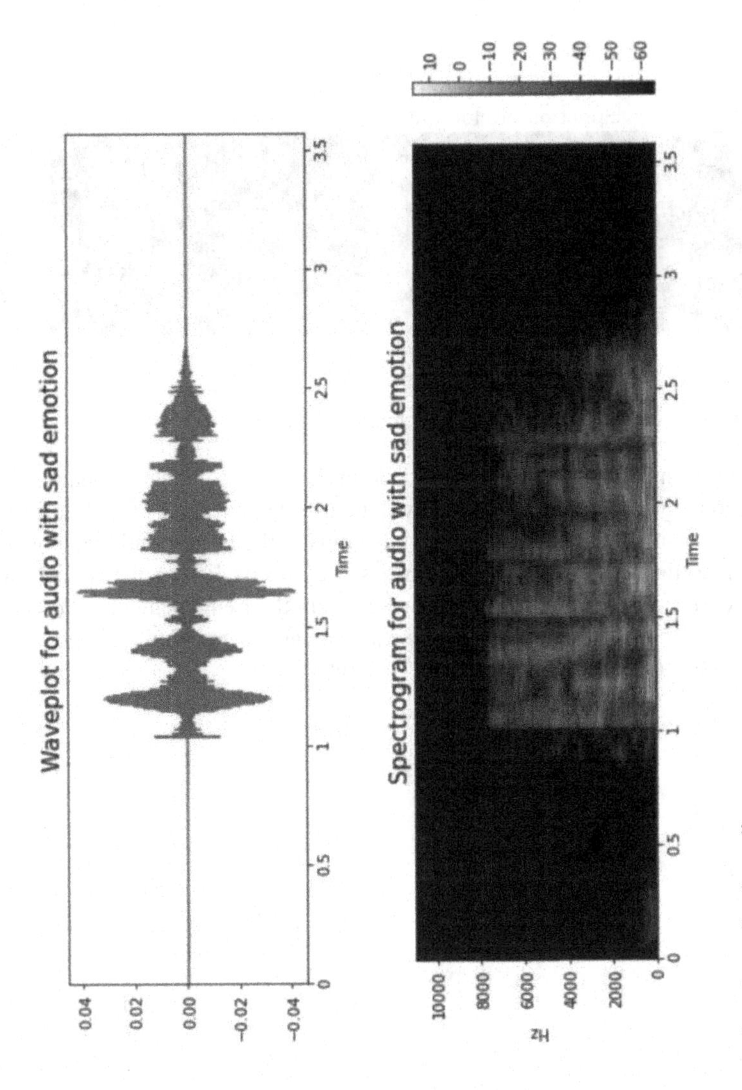

FIGURE 11.4 Learner's sad emotion in audios.

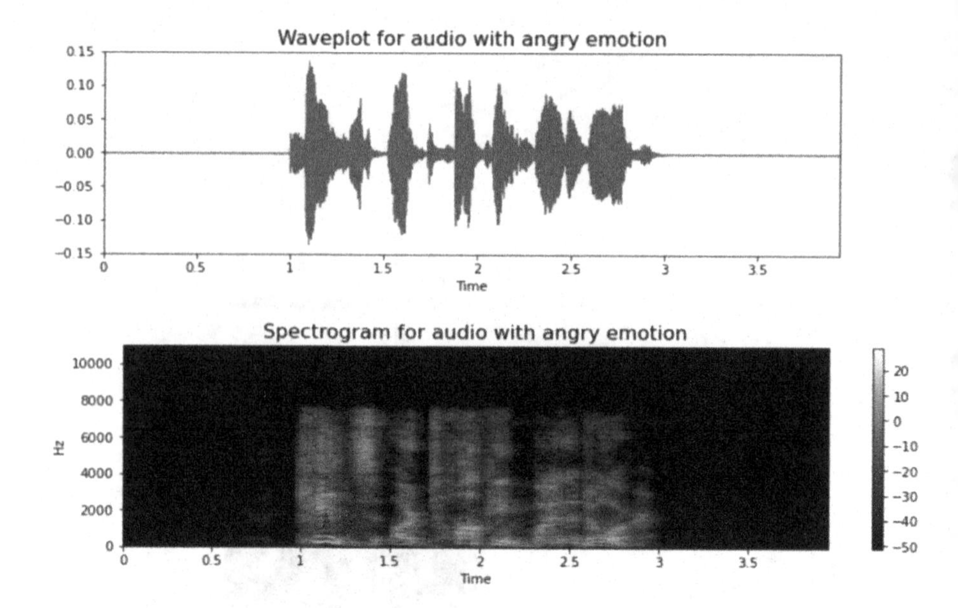

FIGURE 11.5 Learner's confused (angry) emotion in audios.

The audio emotions (sad and confused) from the speech of disabled learners were analyzed to be greater when compared with non-dyslexic learners (see Figure 11.5). They were less involved and not comfortable during the pretest and data collection process. The high count of sad emotion has been derived from the speech of the dyslexic learners. The same has been verified from their engagement data recorded throughout the pretest. Figure 11.6 represents the engagement and motivation of the dyslexic vs non-dyslexic learners during the pretest.

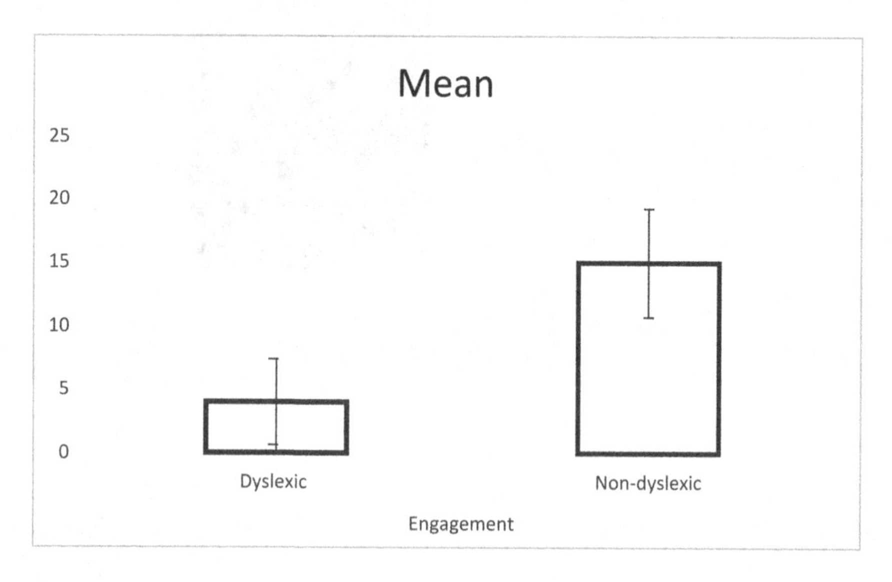

FIGURE 11.6 Learners' feedback analysis for learner motivation and engagement.

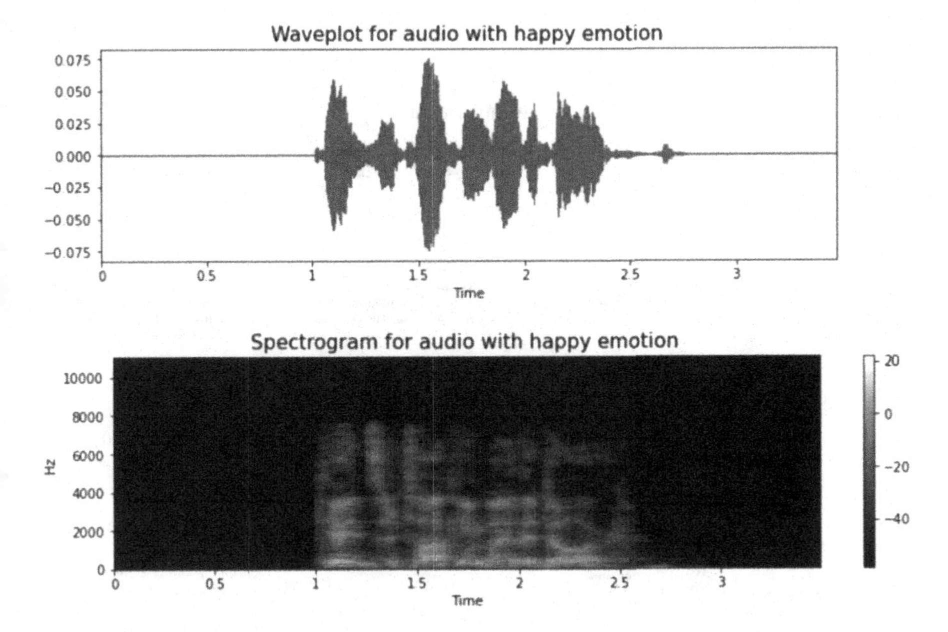

FIGURE 11.7 Learner's excited (happy) emotion in audios.

In the case of non-dyslexic learners, the greatest amount of emotion found in recorded speech was excited emotion. The same was verified from the mean obtained of the learners' feedback data recorded during the pretest (Figure 11.7). The sentiment or emotions analysis from the speech of the learners is further used to determine the problematic skills and adapt the learning content of the learners according to their interest. The pretest questions were mapped to the problematic skills such as phonological awareness, random naming, non-sense words, literacy, reading fluency, etc. If a learner pitch was classified under high (sad or confused) emotion with respect to any of these questions, then that problematic skill is mapped to the learner's profile (please refer to Figure 11.8).

Evaluation of the Observed Emotions of the Dyslexic Learners

In several studies MLP has been considered as a suitable model for classification because it can perform better with complex datasets than KNN or SVM algorithms. The extracted features are provided as input the MLP classifier. Initially, the dataset is split into test and train data. Afterward the trained data are used for classifying the raw dataset. The MLP classification for learners' emotions and its accuracy has been discussed in this section. MLP classifies and contains many hidden layers, but only 100 hidden layers were used to train the data with a maximum of 400 iterations. These hidden layers have improved the accuracy and efficiency of the classification effectively. This MLP-based emotion classification classifies the learners' emotions during pretest as excited, sad, and confused.

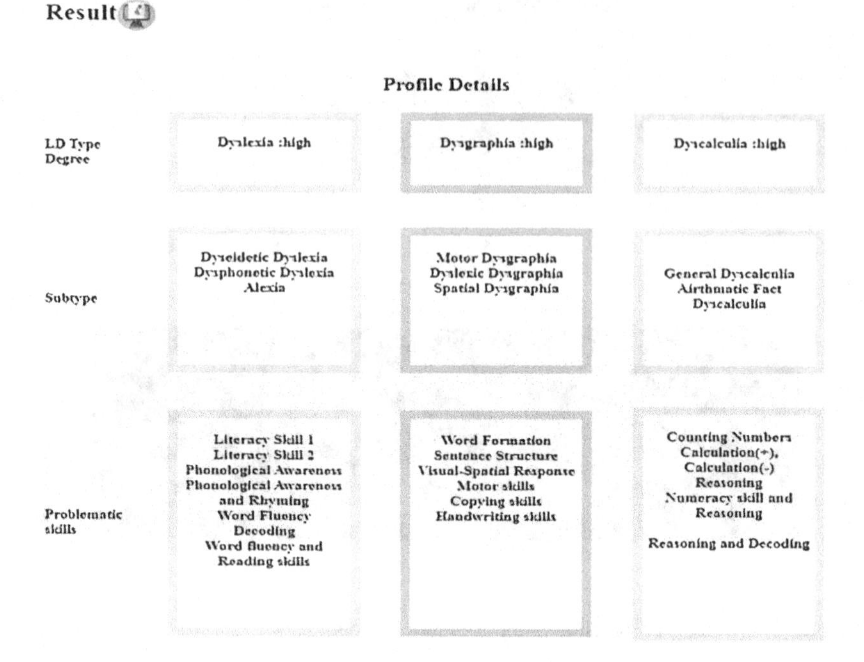

Result
Profile Details
LD Type Degree
Dyslexia :high
Dysgraphia :high
Dyscalculia :high
Subtype
Dyseidetic Dyslexia
Dysphonetic Dyslexia
Alexia
Motor Dysgraphia
Dyslexic Dysgraphia
Spatial Dysgraphia
General Dyscalculia
Airthmatic Fact
Dyscalculia
Problematic skills
Literacy Skill 1
Literacy Skill 2
Phonological Awareness
Phonological Awareness and Rhyming
Word Fluency
Decoding
Word fluency and Reading skills
Word Formation
Sentence Structure
Visual-Spatial Response
Motor skills
Copying skills
Handwriting skills
Counting Numbers
Calculation(+),
Calculation(-)
Reasoning
Numeracy skill and Reasoning
Reasoning and Decoding

FIGURE 11.8 Learner profile mapped with problematic skills.

The performance of ML models has been evaluated using evaluation metrics in different studies. To know how effectively a model is working its evaluation is crucial. Accuracy, precision, and recall are all evaluation metrics that were utilized before in previous studies. Here, we are using accuracy, precision, recall, and F1 score as evaluation metrics for SER of learning-disabled learners.

These evaluation metrics are obtained and compared for confused, excited, and sad emotions of dyslexic and non-dyslexic learners. The evaluation metrics of the model performance on datasets are given in Tables 11.1–11.3.

TABLE 11.1
Accuracy for Test and Train Dataset

	Test	Train
Accuracy	87.9%	98.8%

TABLE 11.2
Precision, Recall, and F1 Score for Test Dataset

	Excited (%)	Confused (%)	Sad (%)
Precision	0.95	**1.00**	0.65
Recall	0.76	0.95	**1.00**
F1 score	0.84	0.98	0.79

TABLE 11.3
Precision, Recall, and F1 Score for Train Dataset

	Excited (%)	Confused (%)	Sad (%)
Precision	1.00	1.00	0.97
Recall	0.99	0.98	1.00
F1 score	0.99	0.99	0.98

Influence of Emotions on the Learning Skills of Dyslexic Learners

Emotions considered in this study for the learner's evaluation are excited, sad, and confused. These three emotions have been determined from the speech of the learners. As a result, the test data has yielded 87.9% accuracy and 98.8% obtained on the train dataset. The evaluation metrics for excited, confused, and sad emotions are calculated for both test and train data. It has been identified in this study that most of the learners with dyslexia did not have excited expression when asked to solve the pretest questionnaires to analyze their skills. This is because they have to put in extra effort compared with non-learning-disabled learners in analyzing questions and providing correct responses. They have given a delayed response for each question compared with the non-learning-disabled learners.

CONCLUSION

The advancements in the domain of AI and ML are leading the way to new technologies. Many automated devices have been invented to support disabled learners in previous research, but for learning-disabled learners extensive research is required to aid in their learning process. Recognizing words is not enough to improve the learning of dyslexic learners; therefore, emotions were comprehended through speech in the proposed study to identify the learner's engagement, motivation, and system usability. This model performance can be improved in terms of efficiency, accuracy, and usability, and by comparing the accuracies with different classification models on the same dataset. Additional to the emotions, the model can be extended to recognize feelings such as depression and mood changes of learning-disabled learners. Also, these systems could be utilized by therapists and parents to monitor the mood swings of dyslexic learners.

REFERENCES

Abbaschian, B. J., Sierra-Sosa, D., & Elmaghraby, A. (2021). Deep Learning Techniques for Speech Emotion Recognition, from Databases to Models. *Sensors, 21*(4), 1249. http://dx.doi.org/10.3390/s21041249

Byun, S.-W., & Lee, S.-P. (2021). A Study on a Speech Emotion Recognition System With Effective Acoustic Features Using Deep Learning Algorithms. *Applied Sciences, 11*(4), 1890. http://dx.doi.org/10.3390/app11041890

Cave, R., & Bloch, S. (2021). The Use of Speech Recognition Technology by People Living With Amyotrophic Lateral Sclerosis: a Scoping Review. *Disability and Rehabilitation: Assistive Technology*, 1–13. doi: 10.1080/17483107.2021.1974961

Dąbrowska, E. (2015). What Exactly Is Universal Grammar, and Has Anyone Seen It? *Frontiers in Psychology, 6*. doi: 10.3389/fpsyg.2015.00852

Daneshfar, F., Kabudian, S. J., & Neekabadi, A. (2020). Speech Emotion Recognition Using Hybrid Spectral-Prosodic Features of Speech signal/glottal Waveform, Metaheuristic-Based Dimensionality Reduction, and Gaussian Elliptical Basis Function Network Classifier. *Applied Acoustics, 166*, 107360. doi:10.1016/j.apacoust.2020.107360

Dutt, S., Ahuja, N. J., & Kumar, M. (2021). An Intelligent Tutoring System Architecture Based on Fuzzy Neural Network (FNN) for Special Education of Learning Disabled Learners. *Education and Information Technologies,* 27, 2613–2633. doi: 10.1007/s10639-021-10713-x

Eginli, A., & Özmelek Taş, N. (2018). Interpersonal Communication in Social Networking Sites: An Investigation in the Framework of Uses and Gratification Theory. *Online Journal of Communication and Media Technologies. 8*. doi: 10.12973/ojcmt/2355

El Hammoumi, O., et al. Emotion Recognition in E-learning Systems. 6th International Conference on Multimedia Computing and Systems (ICMCS). IEEE, 2018.

Habib, M. (2021). The Neurological Basis of Developmental Dyslexia and Related Disorders: A Reappraisal of the Temporal Hypothesis, Twenty Years on. *Brain Sciences, 11*(6), 708. https://doi.org/10.3390/brainsci11060708

Hantke, S., Sagha, H., Cummins, N., & Schuller, B. (2017). Emotional Speech of Mentally and Physically Disabled Individuals: Introducing the EmotAsS Database and First Findings. Proc. Interspeech 2017, 3137–3141. doi: 10.21437/Interspeech.2017-409.

Jacob, U. S., Pillay, J., & Oyefeso, E. O. (2021). Attention Span of Children With Mild Intellectual Disability: Does Music Therapy and Pictorial Illustration Play Any Significant Role? *Frontiers in Psychology, 12*, 677703. https://doi.org/10.3389/fpsyg.2021.677703

Kapoor, K. K., Tamilmani, K., & Rana, N. P. et al. (2018). Advances in Social Media Research: Past, Present and Future. *Information Systems Frontiers, 20*, 531–558. https://doi.org/10.1007/s10796-017-9810-y

Matre, M. E., & Cameron, D. L. (2022). A Scoping Review on the Use of Speech-to-Text Technology for Adolescents With Learning Difficulties in Secondary Education, *Disability and Rehabilitation: Assistive Technology,* doi: 10.1080/17483107.2022.2149865

Sarker, I. H. (2021). Machine Learning: Algorithms, Real-World Applications and Research Directions. *SN Computer Science, 2*, 160. https://doi.org/10.1007/s42979-021-00592-x

Snowling, M. J., & Melby-Lervåg, M. (2016). Oral Language Deficits in Familial Dyslexia: A Meta-Analysis and Review. *Psychological Bulletin, 142*(5), 498–545. https://doi.org/10.1037/bul0000037

Song, Y., & Wei, Z. (2021). Inferring Association Between Alcohol Addiction and Defendant's Emotion Based on Sound at Court. *Frontiers in Psychology, 12*, 669780. https://doi.org/10.3389/fpsyg.2021.669780

12 An Improved Convolutional Neural Network-Based Detection Framework for COVID-19 Omicron and Delta Variants Employing CT Scans

Sunil Kumar, Abhishek Dwivedi, and Shekhar Verma
School of Engineering and Technology (UIET), CSJM University, Kanpur, UP, India

Abhishek Kumar Mishra
IFTM University, Moradabad, UP, India

INTRODUCTION

The new strains of the SARS-CoV-2 virus that produce COVID-19 are the Omicron and Delta variants. In November 2021, South Africa discovered the Omicron variety, while India discovered the Delta variant. Both of these variants have genetic mutations that differ from the previously identified variants of the virus, such as Alpha and Beta. It has been discovered that the Omicron variant of the virus carries a spike protein that has a wide diversity of mutations. The Delta variant also has several mutations, including a mutation in the spike protein, which is thought to contribute to its higher transmissibility and the emergence of new variants. It is important to note that the Omicron and Delta variants have spread rapidly and have become the dominant strains of the virus in many countries [1]. As a result, public health officials and experts are closely monitoring their spread and impact on the ongoing COVID-19 pandemic. Both of these variants have been found to have several

DOI: 10.1201/9781003388319-12

mutations compared with the original virus, which can affect its transmissibility, severity, and response to treatments. For example, the Delta variant is more easily transmitted than previous strains of the virus, and studies have suggested that some people who have recovered from COVID-19 or have been vaccinated may still be susceptible to the Delta variant. It is important to continue monitoring the spread of these variants and their impact [2].

COVID-19, including the new variant known as Omicron, is typically detected through real-time reverse transcription polymerase chain reaction (RT-PCR) tests. This test looks for the presence of the virus' genetic material in a sample taken from the patient, usually from the nose or throat. The sample is collected using a swab and then transported to a laboratory for testing. In the laboratory, the RNA from the sample is reverse-transcribed into copy DNA (cDNA) and then amplified using PCR. If the virus is present in the sample, the amplification of its genetic material will occur, and this can be detected through various methods, such as gel electrophoresis or fluorescent detection. It is crucial to highlight that RT-PCR is one of the numerous kinds of procedures performed to identify COVID-19. There are also rapid antigen tests and antibody tests that can be used to detect the virus, but these tests have different purposes and limitations compared with RT-PCR tests [3].

X-rays, on the other hand, can show the signs of lung damage caused by COVID-19, but cannot detect the virus itself. In x-rays of COVID-19 patients, one can see characteristic patterns of lung involvement, such as increased opacity and decreased lung markings, which are suggestive of pneumonia. However, similar results may also be found in other illnesses such as bacterial pneumonia; therefore, they cannot be used to diagnose COVID-19. It is important to keep in mind that x-rays and other imaging tests, such as computed tomography (CT) scans, are typically employed to assess the severity of lung involvement in individuals with suspected or confirmed COVID-19, rather than to diagnose the virus itself [4]. A CT scan can detect abnormalities in the lungs caused by infections. These variants can cause inflammation and fluid buildup in the lungs, which can be seen on a CT scan as "ground-glass" opacities or areas of hazy, increased density. However, a CT scan alone is not sufficient to diagnose COVID-19 or determine which variant is present. Therefore, CT scans are typically employed as a first-line and important diagnostic tool for COVID-19 detection, but they may be used in combination with other diagnostic procedures to assess the extent of an infection or monitor the progression of lung disease. CT scan is an imaging method that creates collective and comprehensive images of inner organs, bones, and muscles using x-rays [5]. It can detect changes in the lungs caused by respiratory diseases such as COVID-19, including the Omicron and Delta variants. The acronym CNN refers to the sort of technique used in machine learning (ML) known as convolutional neural networks, which can be trained to identify patterns in images. It is a subset of deep learning, a form of artificial intelligence that uses layers of algorithms to recognize and classify patterns in data. CT scans can be analyzed using CNN algorithms to detect the presence of the virus. By training the network on a huge database of CT scans, it can learn to identify the unique patterns and changes associated with Omicron and Delta variants [6].

In essence, the CT scan is used to capture the necessary images, whereas the CNN is used to analyze and interpret those images, providing a more accurate and rapid diagnosis of the virus. Overall, the combination of CT scan and CNN can be a powerful tool in the detection and monitoring, allowing for earlier diagnosis, treatment, and prevention of further spread.

BACKGROUND

Researchers have been looking at a range of diagnostic procedures for Omicron and Delta variants of COVID-19, and one of those ways is CT scans. This research has concentrated on CT scans. Researchers who investigated previous research published mostly on the identification of COVID-19 Omicron and Delta variants discovered so the authors mostly employed CNN in their research.

Gozes and associates centered their efforts on developing a model that can identify COVID-19 in CT images using a small amount of data. The algorithm provides a corona score when a case is positive, indicating the severity of the disease. They suggested a two-dimensional (2D) processing method for per-slice CT scan analysis for each patient, which was done in step one. The 2D region of interest (ROI) segmentation was done to obtain the lungs. A 2D ROI classifier classifies segments as healthy or abnormal in step two. A multiscale model Grad-Cam is used to create a localization map in step three. All segment localization maps are combined in the fourth phase. Corona score biomarkers perform three-dimensional (3D) volumetric scoring in step five. The score indicates sickness severity in step six [7]. Using CT scan images and transfer learning, Ahuja and researchers developed a method to detect COVID-19 infections. They came up with a step-by-step way to find criminals by using the following strategies. In the first step, data was added. In the second step, a CNN model that had already been trained was used. In the third step, CT scan image anomaly localization was done [8]. Silva et al. proposed an approach based on voting for screening COVID-19. Using a voting mechanism, photos are classified into several categories. There is a cross-dataset evaluation available to evaluate the robustness of models utilizing data from diverse distributions [9]. CovidDWNet employs feature reuse residual blocks and depthwise dilated convolutional components. Both components were convolutional. Gradient boosting was used to estimate CovidDWNet-generated feature maps. The CovidDWNet + GB architecture improved CT scan efficiency by over 7% and x-ray imaging by 4% [10]. To take full advantage of the local and global feature extraction benefits offered by transformer for COVID-19 CT images, a parallelism two-branch model called Trans-CNN Net is based on the transformer component as well as the CNN component. Cross-fusion is employed to create a parallelism two-branch feature merger framework that can extract features of varied sizes. When used as a benchmark for testing network classification, the COVIDx-CT dataset demonstrates an accuracy of 96.7%, which is greater than ResNet152, and transformer networks (Deit-B) at 95.2% and 75%, respectively [11]. Two different CNN models were developed by Ghaderzadeh et al. using transfer learning and parameter optimization. In the test dataset, the first phase of the suggested framework had values for a sensitivity of 0.99, a specificity of 0.9986, and an accuracy of 0.988. Phase two of the framework had values for a sensitivity of 0.99, a specificity of 0.9976, and an accuracy of 0.997. The framework reliably categorized CT scans and x-ray images as either containing or not containing COVID-19 [12].

MATERIAL AND METHODS

DATASET

The CT scan datasets that were used to train the CNN models were sourced from Kaggle (Kaggle.com/datasets/mohammadamireshraghi/covid19-omicron-and-delta-variant-ct-scan-dataset) and used for the investigation. The total number of COVID-19 CT scans is 12,231, whereas the total number of scans that do not include COVID-19 is 2251. The instances of the CT scans are presented in Figure 12.1 [13].

DATA PREPARATION

It has been shown via research that the employment of machine algorithms to extract hidden patterns from medical pictures is not primarily attributable to CNN models. Rather, a significant portion of this success may be attributed to the use of proper image preprocessing methods. Images of the dataset were obtained from the Kaggle website and saved in the DICOM format. It is then changed to a grayscale format, which only has one gray channel. CT scan pictures sized 224, 224, and 3 were resized to be compatible with the standard model. The data normalization phase, which is one of the most significant processes in preprocessing, shows a key role not only in raising the speed of picture convergence but also in increasing the pace of training. Following the calculation of the global average and the standard deviation of the pixel surface for each of the photos in the set, the values were then normalized. First, each image's pixel values were adjusted to fall within the range 0–255, and then they were scaled again by converting them to the interval [0,1] [14].

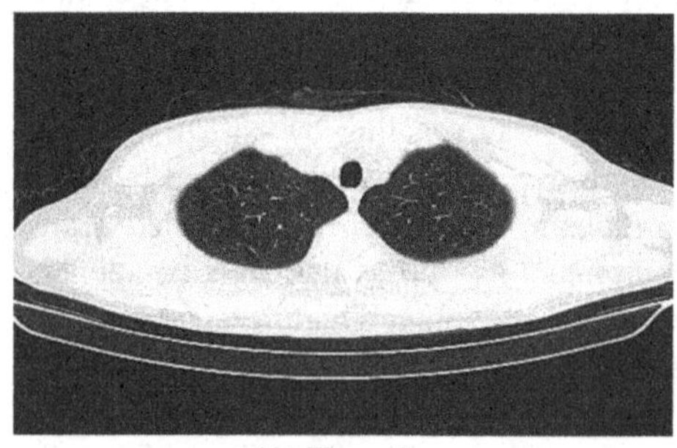

(A) Normal

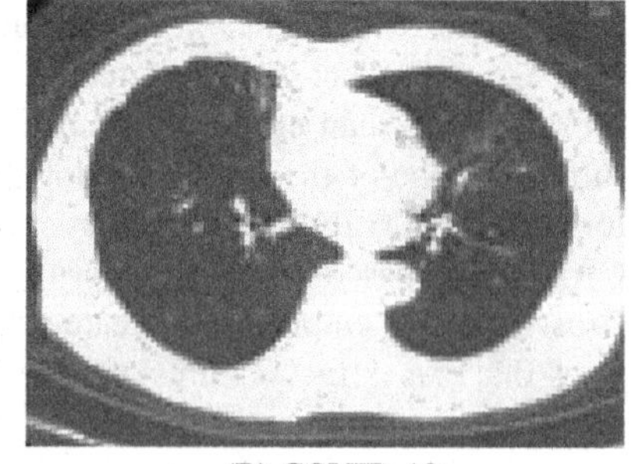

(B) COVID-19

FIGURE 12.1 Instances of CT scans [13].

Feature Extraction and Relevant Feature Selection

Extracting features is one of the most important steps in the process of putting classification algorithms into action. When it comes to the subject of illness diagnosis, this activity has to be carried out with a heightened degree of sensitivity to accurately extract and present the features that identify the irregularities [15]. Following an investigation into the various techniques for the extraction of image characteristics, the authors of this study concluded that a certain form of the pre-trained network was able to abstract the features in a manner that was more effective for each distinct kind of research dataset. The process of feature extraction and the selection of relevant features is carried out in an automated process at CNN [16].

Transfer Learning

Transfer learning involve the process in which an already-trained neural network is employed as a point of entry to solve a different set of problems. In the context of CNNs, transfer learning involves using a pre-trained CNN as an extractor of features and adding a new set of layers on top to perform a new classification task. The pre-trained CNN has already learned to recognize general features in images, such as edges, corners, and textures, and these features can be leveraged to solve a new task with less training data. Pre-trained CNNs are often learned on ImageNet, which contains millions of labeled images. By using transfer learning, we can save time and computational resources that would otherwise be required to train a CNN from scratch. Moreover, transfer learning can improve the accuracy of the new CNN, especially when the size of the new dataset is small [17].

An ML technique permits a pre-trained neural network to be used for a different task than it was originally trained on. This can be especially useful for image classification tasks using CNNs. Here are five popular CNN transfer learning models that are commonly used:

VGG16: The VGG16 model is a deep neural network with 16 layers with over 138 million parameters. There are 13 convolutional layers and 3 fully connected layers in this CNN architecture. This was trained to classify images using the ImageNet dataset, and it achieved very high accuracy. The pre-trained VGG16 model using ImageNet is commonly employed for image classification tasks in transfer learning [18].

ResNet50: The ResNet50 model is a neural network with 50 layers. It was conceived as a solution to the issue of disappearing gradients in deep neural networks, which needed to be addressed. The ResNet50 model is often employed in transfer learning for image classification tasks. ResNet is a deep CNN architecture that consists of residual blocks, which help to improve the training of deep neural networks. ResNet is known for its ability to handle large datasets and deep neural networks [19].

Inceptionv3: The Inceptionv3 model is a deep neural network with 48 layers. It is commonly used for image classification tasks in transfer learning. Inception is a deep CNN that was designed to address the issue of feature redundancy in traditional CNNs. It consists of multiple paths of different convolutions, and each path learns a different type of feature representation [20].

Densenet-121: It is based on the idea of densely connecting each layer to every other layer in a feedforward manner, as opposed to traditional approach of stacking layers one after the other. This leads to a very deep network with a large number of parameters, but also makes it more efficient and easier to train than other deep networks. The dense connections in Densenet-121 allow for better feature propagation through the network, which helps mitigate the vanishing gradient problem that can occur in very deep networks [21].

Improved CNN Sequential Architecture

A sequential architecture is a category of neural network used for the classification of an image and object recognition tasks. The architecture is called "sequential" because it consists of a sequence of layers that are arranged in a specific order. As can be demonstrated in Figure 12.2, the built CNN sequential architecture made use of a 13-layer CNN model, which is a major improvement over the CNN models that were described earlier. The architecture consists of an input layer with a 224 × 224 pixel image size, three Conv2D layers using the ReLU activation

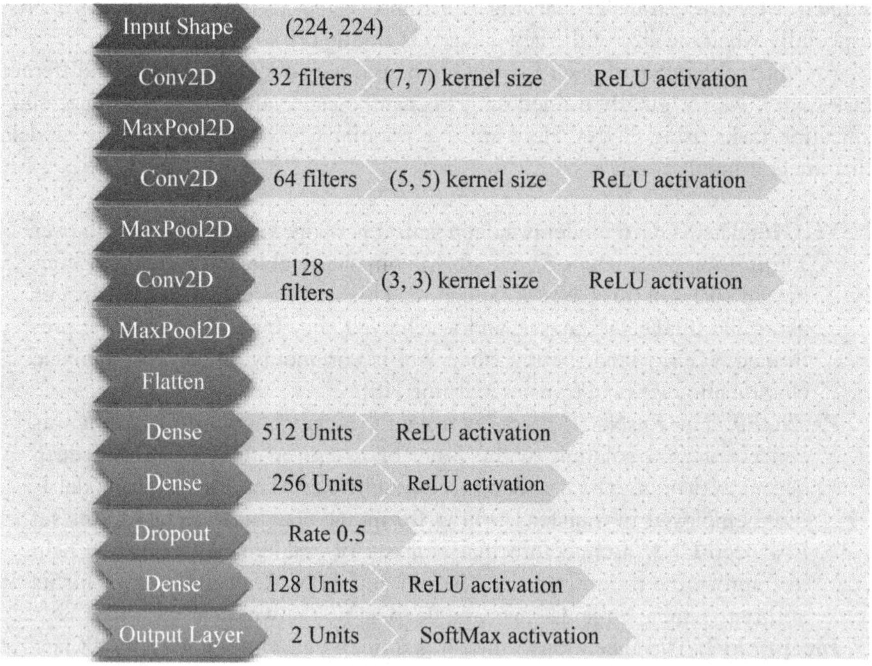

Input Shape (224, 224)
Conv2D 32 filters (7, 7) kernel size ReLU activation
MaxPool2D
Conv2D 64 filters (5, 5) kernel size ReLU activation
MaxPool2D
Conv2D 128 filters (3, 3) kernel size ReLU activation
MaxPool2D
Flatten
Dense 512 Units ReLU activation
Dense 256 Units ReLU activation
Dropout Rate 0.5
Dense 128 Units ReLU activation
Output Layer 2 Units SoftMax activation

FIGURE 12.2 Improved sequential CNN architecture.

function, three Maxpool2D layers for maximum pooling, one flatten layer, three dense layers using the ReLU activation function, a dropout layer, and an output layer using the softmax function.

- **Input Layer:** The input layer of a sequential CNN architecture is a convolutional layer, which applies a set of filters to the input image for the extraction of features.
- **Conv2D:** In our method, we employed a total of 32 filters, 64 filters, and 128 filters for the first, second, and third convolutional layers, respectively, with kernel sizes of 7×7, 5×5, and 3×3, respectively. These filters slide over the image and perform convolutional operations, which involve multiplying the values of the filters with the corresponding values of the input image and summing the results [18]. An activation function that is often used in deep learning neural networks and convolutional operations is referred to as the Rectified Linear Unit, more generally abbreviated as ReLU for short.
- **Maxpool2D:** The convolutional layer's output is transferred to a pooling layer, which downsamples it to minimize feature map dimensionality. This helps to lessen the number of features in the model and prevent overfitting. It is a form of downsampling approach that is utilized to reduce the amount of space that the feature maps, which are created using Conv2D, take up in total [19]. The input feature map is segmented into many tiny, non-overlapping sections for the max-pooling algorithm, and then the maximum value from each of those regions is extracted.
- **Flatten:** A CNN's flatten layer flattens convolutional layer output into a vector. The completely linked layers get this vector. This layer reduces input dimensionality for usage by fully linked layers [20].
- **Dense:** CNN dense layers have every neuron linked to every other neuron in the preceding layer. There is a regular densely connected neural network layer. The dense layer classifies convolutional layer features. The improved sequential architecture made use of a highly connected layer that had 512 neuron units for the dense first layer, 256 neuron units for the dense second layer, and 128 neuron units for the dense third layer [18, 20].
- **Dropout:** CNN regularization uses dropout. It ignores randomly picked neurons during training, and they are randomly "dropped." Their involvement in the actions of downstream neurons is momentarily eliminated while they are in a forward phase, and during a backward pass, weight updates are not applied to the neurons while they are in a backward phase. Neuron weights become less important to the network. This produces a network that generalizes well and doesn't overfit training data. In the course of our investigation, the dropout rate is 0.5%.
- **Output:** The outcome of the preceding layer is then passed to one or more fully connected layers, which perform classification by applying weights to the extracted features, producing a prediction. Typically, a softmax layer constitutes the very last layer in a network, which normalizes the output of the preceding layer and produces a probability distribution over the output classes [21].

TABLE 12.1
Hyperparameters for the Improved CNN Sequential Model

Hyperparameter	Instance
Optimizer	– Adam
Learning rate	– 0.001
Loss function	– Mean-squared logarithmic error loss
Batch size	– 32
Epochs	– 200

The provided model was trained using 32 batches and 200 epochs. The framework was learned with the assistance of an Adam optimizer, which has a learning rate of 0.001 and a mean squared logarithmic error loss as its training parameters. Table 12.1 contains the settings for several hyperparameters, including the optimizer, the learning rate, the loss function, the batch size, and the epochs.

Performance Metrics

The measures that are used to evaluate how well a CNN model works are referred to as performance metrics. The evaluation of categorization models is the primary use for these measures. There are several examples, including accuracy, precision, sensitivity, specificity, and F1 scores. The accuracy and F1 scores were given the most priority in the study because they are the ones that are used the most often and are reliable [22].

RESULTS AND DISCUSSION

The patient's CT scans are then analyzed using VGG16, ResNet50, Inceptionv3, Densenet-121, and improved sequential CNN architecture. This is done to identify COVID-19 instances based on the concealed and multifaceted extracted and relevant features, which are subsequently classified. The issue is to attempt and test created improved sequential CNN architecture with a comparison to VGG16, ResNet50, Inceptionv3, and Densenet-121, considering that CNN is the method that performs the best when classifying medical images. The hyperparameters were implemented in the preceding section.

In the procedure that was introduced, after examining and assessing the most prevalent pre-trained networks, such as VGG16, ResNet50, Inceptionv3, Densenet-121, and improved sequential CNN architecture, it was determined that improved sequential CNN architecture is an effective transverse architecture in the medical research image featuring engineering as well as classification. This conclusion was reached after improved sequential CNN architecture was contrasted to VGG16, ResNet50, Inceptionv3, and Densenet-121. When classifying the features,

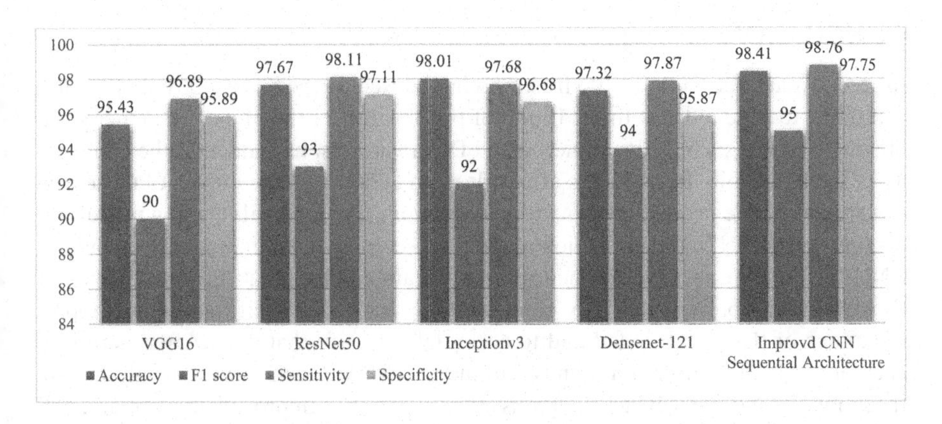

FIGURE 12.3 Various CNNs and improved sequential CNN architecture outcomes.

the layers that make up the classifiers employed the best possible mix of layers. Figure 12.3 shows the results of each CNN algorithm in terms of accuracy, sensitivity, specificity, and F1 score.

Improved sequential CNN architecture accomplished an accuracy of 98.41%, 0.95 F1 score, 98.76% sensitivity, and 97.75% specificity. This indicates that improved sequential CNN architecture was able to correctly detect COVID-19 Omicron and Delta variants efficiently. The results suggest that improved sequential CNN architecture was more accurate and efficient compared with support vector machines (SVMs). It was noticed that the quality and quantity of the data used in the research might potentially cause these conclusions to shift in a different direction. Table 12.2 provides a comparison of the proposed approach to certain alternatives. Researchers preferred accuracy as their primary performance metric, so the comparison is based on accuracy.

TABLE 12.2
Comparison of the Improved Sequential CNN Architecture

Authors	Modality	# Classes	Method	Performance Metrics (%)
Silva et al. [9]	CT scan	2	EfficientCovidNet	Accuracy – 87.60 F1 Score – 86.19
Celik [10]	X-ray and CT scan	4	CovidDWNet	Accuracy – 96.32
Fan et al. [11]	CT scan	3	Proposed Trans-CNN Net, and ResNet-152	Accuracy – 96.70 Specificity – 96.01
Improved sequential CNN architecture	CT scan	2	Sequential CNN architecture	Accuracy – 98.89 F1 Score – 95.00 Sensitivity – 98.76 Specificity – 97.75

CONCLUSION

In recent years, the COVID-19 virus, which is caused by the SARS-CoV-2 virus, has continued to pose a threat to the health of people all over the globe. The emergence of new variants, such as the Omicron and Delta variants, has added to the complexity of the pandemic. Researchers are working to understand the impact of these new variants on public health and provide new solutions like medical imaging modalities to cope with CNN in the prediction of the disease. In sum, an improved sequentially CNN architecture is a type of neural network architecture that uses a sequence of convolutional, pooling, and fully connected layers to train itself hierarchical representations of the data it is fed and to accurately classify that data. After analyzing and comparing the most common pre-trained networks, such as VGG16, ResNet50, Inceptionv3, and Densenet-121, it was established that improved sequential CNN architecture is an excellent transversal design for CT scan images. It is quite promising that the improved sequential CNN architecture achieved an accuracy of 98.41%, a score of 0.95 on the F1 score, 98.76% sensitivity, and 97.75% specificity. In the not too distant future, both the larger dataset and a variety of imaging modalities, such as x-rays, positron emission tomography (PET), and magnetic resonance imaging (MRI), will be investigated.

REFERENCES

1. E. Varea-Jiménez et al., "Comparative severity of COVID-19 cases caused by Alpha, Delta or Omicron SARS-CoV-2 variants and its association with vaccination," Enfermedades infecciosas y microbiologia clinica (English ed.), Feb. 2023, doi: 10.1016/j.eimce.2022.11.021.
2. A. Bolze et al, "Evidence for SARS-CoV-2 Delta and Omicron co-infections and recombination," Med, vol. 3, no. 12, pp. 848–859.e4, 2022, doi: 10.1016/j.medj.2022.10.002.
3. P. Corbisier et al., "A qualitative RT-PCR assay for the specific identification of the SARS-CoV-2 B.1.1.529 (Omicron) variant of concern," Journal of Clinical Virology, vol. 152, p. 105191, Jul. 2022, doi: 10.1016/j.jcv.2022.105191.
4. S. Al-Ahmadi, and F. Mohammad, "Pattern recognition of omicron variants from amalgamated multi-focus EEG signals and X-ray images using deep transfer learning," Egyptian Informatics Journal, Jan. 2023, doi: 10.1016/j.eij.2023.01.001.
5. N. A. Baghdadi, A. Malki, S. F. Abdelaliem, H. Magdy Balaha, M. Badawy, and M. Elhosseini, "An automated diagnosis and classification of COVID-19 from chest CT images using a transfer learning-based convolutional neural network," Computers in Biology and Medicine, vol. 144, p. 105383, May 2022, doi: 10.1016/j.compbiomed.2022.105383.
6. H. C. Reis, and V. Turk, "COVID-DSNet: A novel deep convolutional neural network for detection of coronavirus (SARS-CoV-2) cases from CT and chest X-ray images," Artificial Intelligence in Medicine, vol. 134, p. 102427, Dec. 2022, doi: 10.1016/j.artmed.2022.102427.
7. O. Gozes, M. F. Adar, N. Sagie, and A. Kabakovitch, "A weakly supervised deep learning framework for COVID-19 CT detection and analysis," Thoracic Image Analysis, pp. 84–93, Lima, Peru, October 2020, https://pubmed.ncbi.nlm.nih.gov/36462906/
8. S. Ahuja, B. K. Panigrahi, N. Dey, V. Rajinikanth, and T. K. Gandhi, "Deep transfer learning-based automated detection of COVID-19 from lung CT scan slices," Applied Intelligence, vol. 51, no. 1, pp. 571–585, 2020, doi: 10.1007/s10489-020-01826-w

9. P. Silva et al., "COVID-19 detection in CT images with deep learning: A voting-based scheme and cross-datasets analysis," Informatics in Medicine Unlocked, vol. 20, p. 100427, 2020, doi: 10.1016/j.imu.2020.100427.
10. G. Celik, "Detection of Covid-19 and other pneumonia cases from CT and X-ray chest images using deep learning based on feature reuse residual block and depthwise dilated convolutions neural network," Applied Soft Computing, vol. 133, p. 109906, Jan. 2023, doi: 10.1016/j.asoc.2022.109906.
11. X. Fan, X. Feng, Y. Dong, and H. Hou, "COVID-19 CT image recognition algorithm based on transformer and CNN," Displays, vol. 72, p. 102150, Apr. 2022, doi: 10.1016/j.displa.2022.102150.
12. M. Ghaderzadeh et al., "Efficient framework for detection of COVID-19 Omicron and Delta variants based on two intelligent phases of CNN models," Computational and Mathematical Methods in Medicine, vol. 2022, pp. 1–10, Apr. 2022, doi: 10.1155/2022/4838009.
13. "COVID-19 Omicron and Delta Variant Lung CT Scans," COVID-19 Omicron and Delta Variant Lung CT Scans | Kaggle.com/datasets/mohammadamireshraghi/covid19-omicron-and-delta-variant-ct-scan-dataset
14. K. U. Ahamed et al., "A deep learning approach using effective preprocessing techniques to detect COVID-19 from chest CT-scan and X-ray images," Computers in Biology and Medicine, vol. 139, p. 105014, Dec. 2021, doi: 10.1016/j.compbiomed.2021.105014.
15. M. Kaya, and M. Eris, "D3SENet: A hybrid deep feature extraction network for covid-19 classification using chest X-ray images," Biomedical Signal Processing and Control, vol. 82, p. 104559, Apr. 2023, doi: 10.1016/j.bspc.2022.104559.
16. S. Kumar, H. Kumar, R. Agarwal, and V. K. Pathak, "Human disease prognosis and diagnosis using machine learning," Lecture Notes in Electrical Engineering, pp. 41–53, 2022, doi: 10.1007/978-981-19-0284-0_4.
17. S. Atasever, N. Azginoglu, D. S. Terzi, and R. Terzi, "A comprehensive survey of deep learning research on medical image analysis with focus on transfer learning," Clinical Imaging, vol. 94, pp. 18–41, Feb. 2023, doi: 10.1016/j.clinimag.2022.11.003.
18. K. Simonyan, and A. Zisserman, "Very Deep Convolutional Networks for Large-Scale Image Recognition," arXiv.org, Sep. 04, 2014. https://arxiv.org/abs/1409.1556v6
19. K. He, X. Zhang, S. Ren, and J. Sun, "Deep Residual Learning for Image Recognition," arXiv.org, Dec. 10, 2015. https://arxiv.org/abs/1512.03385v1
20. Z. Liu, C. Yang, J. Huang, S. Liu, Y. Zhuo, and X. Lu, "Deep learning framework based on integration of S-Mask R-CNN and Inception-v3 for ultrasound image-aided diagnosis of prostate cancer," Future Generation Computer Systems, vol. 114, pp. 358–367, Jan. 2021, doi: 10.1016/j.future.2020.08.015.
21. G. Huang, Z. Liu, L. van der Maaten, and Weinberger, Kilian Q, "Densely Connected Convolutional Networks," arXiv.org, 2016. arxiv.org/abs/1608.06993.
22. S. Kumar, and H. Kumar, "Lungcov: A diagnostic framework using machine learning and imaging modality," International Journal on Technical and Physical Problems of Engineering (IJTPE), June 2022. issue 51, vol. 14, Number 2. http://mail.iotpe.com/IJTPE/IJTPE-2022/IJTPE-Issue51-Vol14-No2-Jun2022/23-IJTPE-Issue51-Vol14-No2-Jun2022-pp190-199.pdf

13 A Survey of IoT in Healthcare

Technologies, Applications, and Challenges

S. Subashini, G.K. Kamalam, and P. Vanitha
Kongu Engineering College, Perundurai,
Tamil Nadu, India

INTRODUCTION

In recent years, there have been various life-threating diseases that have created various public health issues like COVID-19. The inability to access various healthcare sectors like hospitals, particularly by the elderly and disabled people, raises the need for remote healthcare monitoring systems. Internet of Things (IoT) is a technology comprised of various wireless technologies and sensors used to gather the data from patient's body remotely. Also, the collected data can be transmitted to the patient's doctors and can be utilized for immediate action and various data analysis. IoT is applied in diverse areas like automated irrigation systems in agriculture, smart homes, environment monitoring, disaster monitoring, etc.

The wireless sensor network (WSN) is comprised of sensor nodes that can sense various data like temperature, moisture, heart rate, etc., and the data can be transmitted to the server through the Internet. Wireless body area network (WBAN) is the technology widely adopted in remote healthcare monitoring systems to collect certain vital parameters from the human body. IoT is the powerful technology that integrates various devices, humans, and things through various smart technologies, such as WSNs and WBANs. It enables remote data collection through sensors, radio-frequency identification (RFID), Bluetooth, and other communication interfaces. IoT can be seen and used as an important technology in remote healthcare management because there is an increase in the need to make healthcare more individualized and economical.

The implementation of IoT in healthcare functions in three phases like tracking of people like doctors and patients and user authentication and automatic data sensing and gathering. Tracking people involves searching the doctors who are all in the nearby vicinity in case of emergency situations in hospitals, searching for Alzheimer's patients, etc. The patient's data stored in the server should be accessed only by authenticated persons like their doctor and family members. This brought about the requirement for appropriate user authentication mechanisms. The WSNs

 DOI: 10.1201/9781003388319-13

and WBANs play a vital role in data sensing and collection, and the gathered data are transmitted through several communication technologies.

This chapter is organized as follows. The first section describes the related work and architecture of IoT in healthcare, which illustrates the key components. The various technologies that are adopted are explained in the next section. IoT is applied diversely in various subfields of healthcare, which is detailed in the following section. Next, the various challenges to be addressed in implementing IoT in healthcare are discussed. Further, the future directions of IoT are briefed and concluded in the last two sections.

IoT IN HEALTHCARE: RELATED WORK

There is a variety of approaches adopted for healthcare-based applications in IoT including sensor-based approaches, communication-based approaches, and other approaches based on resources, security, and applications [1]. The sensor-based approaches involved the usage of wearable sensors that are attached to the human body. Various devices that are connected to the Internet and mobile devices are utilized in association with wearable gadgets. Information is continuously gathered and kept in the cloud server from the mobile device. The health conditions of sportsperson could be continuously monitored based on wearable technology and ensemble Bayesian deep classifier (EBDC) approach [2]. The authors proposed a wearable sensors-based disease prediction combined with IoT and deep learning methodology, which is utilized to train data for an artificial intelligence (AI) device [3].

The communication-based approach involves managing the communication interfaces by adopting various algorithms and technologies. Using various technologies, including smart mobile, RFID, and WSN, a hospital management system based on IoT was proposed that ensures automatic monitoring of patient's data and biomedical devices within the healthcare system [4]. To provide individualized monitoring for covering a large geographic area, mobility, simple installation, and a long battery life, an IoT-based analyzer along with electronic reader, which is enabled through Bluetooth or Lora, was proposed [5]. The authors developed an optimal routing methodology based on the proximity of health centers and the continuous monitoring in crises in addition to choosing an optimal route that enables the visitors to reach faster [6]. A dependable mobile-to-mobile-based IoT system designed for healthcare devices was provided by the authors with a fault-tolerant algorithm that made use of data from different sources like backup, integrity, and daisy chain data [7].

The goal of the application-based technique is to provide systems that could carry out different types of tasks. A healthcare monitoring system adopted using the cloud for predicting various diseases like diabetes along with classification algorithms was adopted by the authors [8]. An application designed for healthcare based on fog computing was proposed to manage hypertension [9]. For healthcare-based IoT systems, the authors suggested a real-time monitoring architecture in which neural network classifiers were adopted in addition to decision trees to report tumor growth with a high degree of predictability [7].

The authors [10] proposed an architecture for IoT-based healthcare applications. The architecture comprises three layers, such as the perception layer, network layer, middle-ware layer and application layer.

- Perception layer: This layer is the bottommost layer; it has the responsibilities of physical layer. There are various sources of data that have to be collected from the patient's body like heart rate, blood pressure, etc., through wearable sensors. The sensed data have to be collected, which is performed in the perception layer along with signaling operations.
- Network layer: The data collected from patients are transmitted to this layer, which can link all smart devices collectively and enable the transmission of healthcare information between them. This layer uses wireless technologies like Wi-Fi, Zigbee, Bluetooth, and infrared to securely transfer health information from patients to the base station.
- Middleware layer: The middleware layer provides services to the requester that are connected to names and addresses. The Healthcare IoT application programmers don't need to pay attention for a particular equipment platform while working with non-homogeneous items. This layer is used to transmit the collected health information from the network layer and the gathered data is stored in the server.
- Application/business layer: This layer hires healthcare providers to analyze and combine the information that has been provided by other tiers. To meet patients' needs, this layer can offer various healthcare services. This layer is in charge of overseeing all operations and healthcare services through the development of well-implemented innovations and technologies, like graphs, business models, and visualized representations.

The architecture of the IoT-based healthcare system is depicted in Figure 13.1, which comprises of key functionalities like data generation, data processing, and information consuming. Data generation involves collecting data from the human body through various sensors and processed using tools for analyzing data and other machine learning algorithms. Data generation is performed through the perception layer and network layer and data processing is performed through the middleware layer. The results and analytics from the data processing phase can be used for any necessary decision-making by healthcare practitioners in the information consuming phase.

IoT IN HEALTHCARE: TECHNOLOGIES ADOPTED

The interconnection of the several elements of the Healthcare IoT network is achieved through various short-range and medium-range communication systems [11]. The medium-range communication technologies typically support communication over a large distance, like communication between a base station and the central node of a BAN. The short-range communication technologies are the protocols used to connect objects within a limited range or within the network ranging between

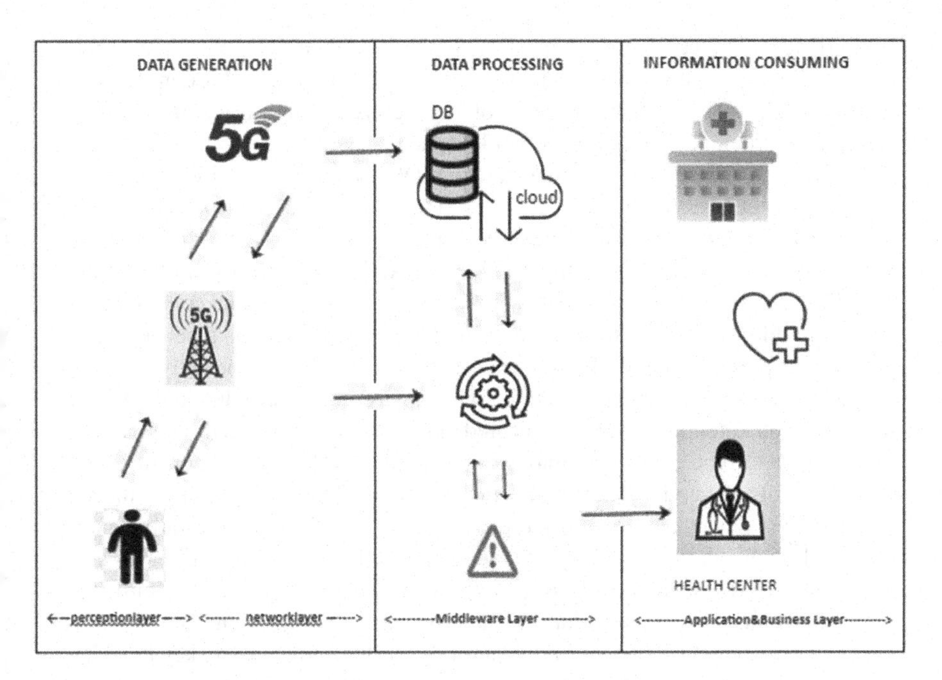

FIGURE 13.1 Architecture of IoT based healthcare.

few centimeters to several meters. Some of the commonly adopted technologies are Bluetooth, RFID, Wi-Fi, etc.

- Bluetooth: This is a short-range wireless communication technique that makes use of UHF radio waves through which two or more medical devices can communicate with one another. It works under the frequency band of 2.4 GHz with a communication range of about 100 m. Data protection is provided through encryption and authentication mechanisms. The significant features of Bluetooth are low cost, energy efficiency, and less interferent data transmission. However, this technology falls short of the requirement when the healthcare application calls for long-distance communication.
- RFID: This is employed for close-proximity communication ranging between 10 cm to 200 m. It recognizes and follows tags attached to objects impulsively using an electromagnetic field. The reader communicates with a tag utilizing radio waves to send or receive data from the object. The RFID node of the IoT system was adopted to monitor the body temperature and the data are transmitted to the server in real time. RFID makes it possible for medical professionals to instantly find and monitor medical equipment. Its key benefit is that it doesn't require an external power supply. It is, however, a very unreliable protocol that could have compatibility concerns when connecting to a smartphone.
- Zigbee: One of the common protocols used to link medical devices and to exchange data among them is called Zigbee. It does, however, have a

greater connection range than Bluetooth devices and supports mesh networking topology. Through mesh topology, it achieves better connectivity and fault tolerance. The benefits of Zigbee are its high transmission rate, power efficiency, and large network capacity.

- Wi-Fi: A wireless local area network (WLAN) called wireless fidelity (Wi-Fi) adheres to the IEEE 802.11 standard. Compared with Bluetooth, it offers a greater communication range at about 70 feet. Wi-Fi creates networks in an ad hoc manner, which makes it easier to use in hospitals. The widespread use of Wi-Fi is due to its simple smartphone compatibility and ability to enable strong control and security. It exhibits a considerably higher power demand, though, and the network operates unevenly.
- Near field communication (NFC): The electromagnetic induction between two loop antennas, which are located close to one another, is the fundamental idea of NFC. This technology is comparable to RFID in that it similarly transmits data via electromagnetic induction. The two operating modes for NFC devices are active and passive. The radiofrequency is only generated by one device in passive mode, whereas the second device serves as a receiver. When in active mode, both devices can simultaneously emit the radiofrequency and send data without pairing. The simplicity of operation and effective wireless communication network of NFC are its key benefits. However, it only works for very close-range communication.
- Satellite: In remote and widely separated geographic places where other communication methods are difficult to access, satellite communication is proven to be more efficient and advantageous. The signals from the land are picked up by the satellite, amplified, and then sent back to Earth. Around the Earth, more than 2000 satellites are in orbit. High-speed data transfer, immediate Internet access, stability, and technological compatibility are all benefits of satellite communication. However, compared with other communication methods, satellite transmission has a high power consumption.

IoT APPLICATIONS IN THE HEALTHCARE SYSTEM

- Remote health care monitoring service: Remote healthcare observing is one of the useful and well-liked uses of the health services offered by the IoT (telehealth). Patients no longer always need to go to the hospital or even the emergency room thanks to the deployment of IoT. In addition to lowering costs and reducing the necessity for hospital visits, this service also enhances patient's life quality by relieving them of the uneasiness of hospital travel. In the case of patients who have trouble walking or using public transportation, even such a small convenience can make a big impact [12].
- Assisted living and elderly care: Given that people are living longer and there are an increasing number of senior people in the world, it is evident that the current health care systems are ill-equipped to care for elderly patients, both monetarily and in terms of physical resources. The bulk of the elderly must be kept under minimal monitoring majority of the time for a variety of factors, largely medical. Elderly individuals' health data can be

continuously monitored thanks to the IoT. Additionally, improved assisted living surroundings with IoT integration will make life for persons with Alzheimer's, Parkinson's, dementia, etc., much more comfortable as well as easier for their families and coworkers.

- Recognizing and treating chronic illnesses: With an extended lifespan, chronic health issues have also become more prevalent. Fortunately, there have been significant advancements in the treatment of these issues, and IoT deployment can have a significant impact on many of them. The introduction of numerous methods of diagnosing and treating chronic diseases can be directly facilitated by wearable technology, next-generation analytical tools, and mobile devices. IoT devices are used by utilities similar to Fitbit to provide long-term, continuous monitoring of individual health. It is simple to provide information to the patient's doctor to aid in the treatment of chronic illnesses. The Health Net Connect-developed diabetic management software, which aims to enhance medical care and lower medical expenses for people with diabetes, is an illustration of such IoT-based applications.
- Personalized medication: New tailored prescriptions and pharmacological treatments are a very interesting trend involving IoT usage in the health industry. Although it may still sound like science fiction, tablets with rice-grain-sized tiny sensors are now able to communicate with an external device worn by patients to monitor dosage and drug usage. This knowledge can also be very helpful to patients by prompting them to print their prescriptions or make a record of the medications they need to take. To track their own performance and change their habits, patients will also have access to all of their personal data through a convenient smartphone app.
- Reducing wait times for emergencies: For those who are awaiting in an emergency room, a few things stand out in particular. In addition to the monetary outlay, emergency department medical procedures can occasionally take hours to perform. The New York Sinai Medical Center has effectively cut the wait periods for patients who require in-patient care by over 50%. The solution, offered by the IoT-based software "AutoBed" from GE Healthcare, checks the tenancy rate among 1200 units and 15 different metric parameters to determine each patient's demands. Despite being a relatively old example, it served as an inspiration for other more contemporary uses at various hospitals. It serves as a good illustration of an IoT-based innovative system that is highly effective.
- Observing hospital staff, equipment, and safety: While keeping track of hospital resources, such as staff, patients, and equipment, is simple for smaller facilities, it has become very difficult for many large city and urban hospitals with tens of thousands of patients and staff, as well as outpatient clinics and various units. To make resource and service tracking easier, many healthcare companies all around the world are turning to IoT and available positioning systems. IoT-based systems offer not only a cheap means to keep an eye on routine hospital activities, but they also offer a cutting-edge product that makes it possible to organize and keep track of these incredibly complex entities.

- Assessment of vital medical equipment: For modern hospitals to function and provide services, a new generation of medical hardware and software is necessary Depending on the situation, this danger may be the difference between life and death. Philips' so-called "e-Alarm System" is an illustration of an all-encompassing IoT system that seeks to address the issue. Philips' new technology provides a proactive alternative to waiting for a medical item to malfunction in a hospital by electronically observing medical equipment and notifying hospital workers of the problems.
- IoT-enabled ambulances: Because they deal with more serious patients and quick decisions are essential to saving the patient's life, ambulance staff have particularly stressful jobs. An IoT-ènabled ambulance is particularly effective because it allows a remote medical team to recommend what has to be done for the patient. As a result, the patient receives prompt attention and efficient treatment in IoT-enabled ambulances. The Life First Emergency Traffic Control algorithm, which modifies the traffic signal design or duration for ambulances and emergency service providers during an emergency, was created by Red Ninja as the first firm.
- Monitoring of glucose: The monitoring of blood pressure, temperature, and glucose levels is a common practice, and COVID-19 has benefited greatly from this. Patients with diabetes may have other metabolic problems, such as persistently elevated blood sugar. Monitoring blood sugar levels helps with daily activities like eating, drinking, and taking medication by revealing the pattern of blood glucose changes. Real-time glucose level monitoring has been proposed, and the data are uploaded to a server and subsequently transmitted to particular doctors for additional processing. For data exchange with the relevant healthcare service provider, the sensors are continuously connected to IPV6. The IoT has made it possible for patients and service providers to exchange data.

CHALLENGES OF IoT IN HEALTHCARE

- Scalability: It is an important factor in the healthcare system that indicates the system's capacity to meet changing demands and adjust to larger-scale changes in the environment. According to the literature review, some of the suggested methods for healthcare-based IoT systems can work on a limited scale. Applications need to be able to integrate with a variety of devices and systems for the architecture to be scalable in case of a sudden rise in the number of linked devices.
- Security: Integrity and security management are important unresolved challenges for accessing and storing data in IoT-based healthcare platforms. Integrity can be characterized as regulating access to credentials, protecting patient and service provider confidentiality, and preventing unauthorized access to any of these things. IoT-based systems for the healthcare industry are developed based on the data collected by IoT nodes. Additionally, as the number of IoT devices connected to the network grows, so does the susceptibility of the vast amounts of data being sent and stored [13].

- Managing resources: The nodes deployed in healthcare-based IoT stations in which processing and storage capacity are limited need efficient resource management techniques. Also, resource management could effectively use a variety of services to generate an effective use of the resources available and provide appropriate services.
- Mobility: It refers to the capacity to leverage network support for patients to connect to the gateway whenever and wherever they choose. Mobility is also required to improve service quality, allow comprehensive access to information regardless of users' locations, and make the network fault-tolerant. A mobility protocol must be reliable to reduce packet losses, network failures in any situation, and end-to-end latencies

FUTURE OF IoT IN HEALTHCARE

Although there has been significant advancement in sensors, there are still no products that can match the precision of hospital-grade devices without sacrificing wearability or energy efficiency. As a result, more research should be done to enhance the quality of these sensors to improve the accuracy, reliability, and comfortability. The development of wearable healthcare systems that rely on the NB-IoT standard would be beneficial in terms of communications standards.

CONCLUSION

IoT provides numerous solutions in the healthcare field ranging from simple models to complex and smart system designs. Various technologies that have been adopted in achieving solutions for IoT-based healthcare systems were RFID, Wi-Fi, NFC, LoRaWAN, etc. The architecture of this system involves the transfer of data from the patient's body to doctors, which is stored in a database server and utilized for various data analyses. Various challenges like resource management, mobility, and scalability are addressed through appropriate methodologies. As it is adopted in healthcare for monitoring various vital parameters, a great deal of attention should be taken in providing the accuracy of the data. As a large amount of data need to be collected, the healthcare system becomes very complex and has to overcome various challenges in data processing. In the future, low power communication standards along with machine learning algorithms could be adopted to overcome the challenges in healthcare systems.

REFERENCES

1. Kashani, M.H., Madanipour, M., Nikravan, M., Asghari, P. and Mahdipour, E., 2021. A systematic review of IoT in healthcare: Applications, techniques, and trends. Journal of Network and Computer Applications 192. https://doi.org/10.1016/j.jnca.2021.103164.
2. Huifeng, W., Kadry, S.N. and Raj, E.D., 2020. Continuous health monitoring of sportsperson using IoT devices based wearable technology. Computer Communications. https://doi.org/10.1016/j.comcom.2020.04.025.
3. Muthu, B., Sivaparthipan, C.B., Manogaran, G., Sundarasekar, R., Kadry, S., Shanthini, A. and Dasel, A., 2020. IOT based wearable sensor for diseases prediction and symptom analysis in healthcare sector. Peer-to-Peer Networking and Applications 13 (6), 2123–2134.

4. Catarinucci, L., De Donno, D., Mainetti, L., Palano, L., Patrono, L., Stefanizzi, M. and Tarricone, L., 2015. An IoT-aware architecture for smart healthcare systems. IEEE Internet of Things Journal 2, 1–1. https://doi.org/10.1109/JIOT.2015.2417684.
5. Catherwood, P.A., Steele, D., Little, M., Mccomb, S. and Mclaughlin, J., 2018. A community based IoT personalized wireless healthcare solution trial. IEEE Journal of Translational Engineering in Health and Medicine 6, 1–13.
6. Almobaideen, W., Krayshan, R., Allan, M. and Saadeh, M., 2017. Internet of things: Geographical routing based on healthcare centers vicinity for mobile smart tourism destination. Technological Forecasting and Social Change 123, 342–350.
7. Suresh, A., Udendhran, R., Balamurgan, M. and Varatharajan, R., 2019. A novel internet of things framework integrated with real time monitoring for intelligent healthcare environment. Journal of Medical Systems 43 (6), 165–175.
8. Verma, P. and Sood, K.S., 2018. Cloud-centric IoT based disease diagnosis healthcare framework. Journal of Parallel and Distributed Computing 116, 27–38. https://doi.org/10.1016/j.jpdc.2017.11.018.
9. Sood, S.K. and Mahajan, I., 2019. IoT-fog based healthcare framework to identify and control hypertension attack. IEEE Internet of Things Journal 6 (2), 1920–1927.
10. Farooq, M.U., Waseem, M., Khairi, A. and Mazhar, S., 2015. A critical analysis on the security concerns of internet of things (IoT). International Journal of Computer Applications 111, 1–6. https://doi.org/10.5120/19547-1280.
11. Alshamrani, M., 2022. IoT and artificial intelligence implementations for remote healthcare monitoring systems: A survey. Journal of King Saud University – Computer and Information Sciences 34, 4687–4701.
12. Ray, P.P., Dash, D. and De, D., 2019. Analysis and monitoring of IoT-assisted human physiological galvanic skin response factor for smart e-healthcare. Sensor Review 39 (4), 525–541.
13. Boussada, R., Hamdane, B., Elhdhili, M. and Saidane, L., 2019. Privacy-preserving aware data transmission for IoT-based E-health. Computer Networks 162 (9), 106866.

Index

S

T

V

W

X

Y

Z

Printed in the United States
by Baker & Taylor Publisher Services